Calculus: The Basics

Calculus: The Basics

Megan Baker

NY RESEARCH
P R E S S

New York

Published by NY Research Press
118-35 Queens Blvd., Suite 400,
Forest Hills, NY 11375, USA
www.nyresearchpress.com

Calculus: The Basics
Megan Baker

International Standard Book Number: 978-1-63238-877-3 (Hardback)

This book contains information obtained from authentic and highly regarded sources. All chapters are published with permission under the Creative Commons Attribution Share Alike License or equivalent. A wide variety of references are listed. Permissions and sources are indicated; for detailed attributions, please refer to the permissions page. Reasonable efforts have been made to publish reliable data and information, but the authors, editors and publisher cannot assume any responsibility for the validity of all materials or the consequences of their use.

Trademark Notice: Registered trademark of products or corporate names are used only for explanation and identification without intent to infringe.

Cataloging-in-Publication Data

Calculus : the basics / Megan Baker.
     p. cm.
Includes bibliographical references and index.
ISBN 978-1-63238-877-3
1. Calculus. 2. Mathematical analysis. 3. Functions. 4. Geometry, Infinitesimal.
I. Baker, Megan.
QA303.2 .C35 2022
515--dc23

Contents

Preface

The mathematical study of continuous change is known as calculus. There are two major divisions of calculus, known as differential calculus and integral calculus. The instantaneous rates of change and the slope of curves is studied under differential calculus. Integral calculus deals with the accumulation of quantities and areas between and under the curves. The link between these two branches is known as the fundamental theorem of calculus. This theorem states that differentiation and integration are inverse operations. Calculus finds its application in every field where a problem can be mathematically modeled and where an optimal solution is sought. Therefore, it is used in all branches of physical science, actuarial science, computer science, statistics, engineering and a variety of other disciplines. The topics covered in this extensive book deal with the core aspects of calculus. It is appropriate for students seeking detailed information in this area as well as for experts. The book will serve as a valuable source of reference for graduate and post graduate students.

A detailed account of the significant topics covered in this book is provided below:

Chapter 1- The branch of mathematics which is involved in the study of continuous change is known as calculus. It has two major branches, namely, differential calculus and integral calculus. This is an introductory chapter which will introduce briefly all the significant aspects of calculus such as limits and continuity, and vector calculus.

Chapter 2- The derivative of a function of a real variable is used to measure the sensitivity to change of the function value in relation to a change in its input value. Some of the concepts studied in relation to derivatives are local extrema of functions and rules for finding derivatives. This chapter closely examines these key concepts of derivatives to provide an extensive understanding of the subject.

Chapter 3- Integral is an operation of calculus which assigns numbers to functions such that it describes displacement, area, volume, and other concepts which occur due to combination of infinitesimal data. The various techniques used within integration are integration by substitution, integration by parts and integration by trogonometric substitution. This chapter discusses in detail these techniques of integration.

Chapter 4- The elements of vector space are known as vectors. There are a number of important areas of study related to vectors such as vector functions, gradient, curl and divergence. The chapter closely examines these key concepts of related to vectors to provide an extensive understanding of the subject.

Chapter 5- The field of calculus uses a number of theorems such as Rolle's Theorem, divergence theorem, gradient theorem, Stokes' theorem, Green's theorem and mean value theorem. The diverse applications of these theorems have been thoroughly discussed in this chapter.

It gives me an immense pleasure to thank our entire team for their efforts. Finally in the end, I would like to thank my family and colleagues who have been a great source of inspiration and support.

Megan Baker

Introduction to Calculus

The branch of mathematics which is involved in the study of continuous change is known as calculus. It has two major branches, namely, differential calculus and integral calculus. This is an introductory chapter which will introduce briefly all the significant aspects of calculus such as limits and continuity, and vector calculus.

Calculus is a branch of Mathematics that deals with the study of limits, functions, derivatives, integrals and infinite series. The subject comes under the most important branches of applied Mathematics, and it serves as the basis for all the advanced mathematics calculations and engineering applications.

Categories of Calculus

There are two major categories of Calculus:

1. Differential Calculus

2. Integral Calculus

Functions

A function is defined by three elements:

1. A domain: A subset of $\mathbb{R}$. The numbers which may be "fed into the machine".

2. A range: Another subset of $\mathbb{R}$. Numbers that may be "emitted by the machine". We do not exclude the possibility that some of these numbers may never be returned. We only require that every number returned by the function belongs to its range.

3. A transformation rule: The crucial point is that to every number in its domain corresponds one and only one number in its range.

We normally denote functions by letters, like we do for real numbers (and for any other mathematical entity). To avoid ambiguities, the function has to be defined properly. For example, we may denote a function by the letter f. If $A \subseteq \mathbb{R}$ is its domain, and $B \subseteq \mathbb{R}$ is its range, we write f: A → B (f maps the set A into the set B). The transformation rule has to specify what number in B is assigned by the function to each number $x \in A$. We denote the assignment by $f(x)$ (the function f evaluated at x). That the assignment rule is "assign $f(x)$ to x" is denoted by $f: x \rightarrow f(x)$ (pronounced "f maps x to $f(x)$").

Let $f: A \to B$ be a function. Its image is the subset of B of numbers that are actually assigned by the function. That is:

$$\text{Image}\left(f\right) = \left\{y \, B : x \, A : f\left(x\right) = y\right\}.$$

The function f is said to be onto B if B is its image. f is said to be one-to-one if to each number in its image corresponds a unique number in its domain, i.e.,

$$\left(\forall y \in \text{Image}\left(f\right)\right)\left(\exists! x \in A\right) : \left(f\left(x\right) = y\right).$$

Examples: A function that assigns to every real number its square. If we denote the function by f, then,

$$f : \mathbb{R} \to \mathbb{R} \quad \text{and} \quad f : x \to x^2$$

We may also write $f\left(x\right) = x^2$.

We should not say, however, that "the function f is x^2". In particular, we may use any letter other than x as an argument for f. Thus, the functions $f : \mathbb{R} \to \mathbb{R}$, defined by the transformation rules: $f\left(x\right) = x^2$, $f\left(t\right) = t^2$, $f\left(\alpha\right) = \alpha^2$ and $f\left(\xi\right) = \xi^2$ are identical.

Also, it turns out that the function f only returns non-negative numbers. There is however nothing wrong with the definition of the range as the whole of $\mathbb{R}$. We could limit the range to be the set $[0, \infty)$, but not to the set.

A function that assigns to every $w \neq \pm 1$ the number $(w^3 + 3w + 5)/(w^2 - 1)$. If we denote this function by g, then,

$$g : \mathbb{R} \setminus \left\{\pm 1\right\} \to \mathbb{R} \quad \text{and} \quad g : w \to \frac{w^3 + 3w + 5}{w^2 - 1}.$$

A function that assigns to every $-17 \leq x \leq \pi/3$ its square. This function differs from the function in the first example because the two functions do not have the same domain (different "syntax" but same "routine").

A function that assigns to every real number the value zero if it is irrational and one if it is rational. This function is known as the Dirichlet function. We have $f : \mathbb{R} \to \left\{0, 1\right\}$, with,

$$f : x \mapsto \begin{cases} 0 & x \text{ is irrational} \\ 1 & x \text{ is rational.} \end{cases}$$

A function defined on the domain,

$$A = \left\{2, 17, \pi^2 /17, 36/\pi\right\} \cup \left\{a + b\sqrt{2} : a, b \in \mathbb{Q}\right\},$$

such that,

$$x \mapsto \begin{cases} 5 & x = 2 \\ 36/\pi & x = 17 \\ 28 & x = \pi^2/17 \text{ or } 36/\pi \\ 16 & \text{Otherwise.} \end{cases}$$

The range may be taken to be $\mathbb{R}$.

A function defined on $\mathbb{R} \setminus \mathbb{Q}$ (the irrational numbers), which assigns to x the number of 7's in its decimal expansion, if this number is finite. If this number is infinite, then it returns $-\pi$. This example differs from the previous ones in that we do not have an assignment rule in closed form (how the heck do we compute $f(x)$?). Nevertheless it provides a legal assignment rule.

For every $n \in \mathbb{N}$ we may define the nth power function $f_n : \mathbb{R} \to \mathbb{R}$, by $f_n : \cdot x \to x^n$. Here again, we will avoid referring to "the function x^n". The function $f_1 : x \mapsto x$ is known as the identity function, often denoted by Id, namely:

$$Id : \mathbb{R} \to \mathbb{R}, \ Id : x \mapsto x.$$

There are many functions that you all know since high school, such as the sine, the cosine, the exponential, and the logarithm. These functions require a careful, sometimes complicated, definition.

Given several functions, they can be combined together to form new functions. For example, functions form a vector space over the reals. Let $f : A \to \mathbb{R}$ and $g : B \to \mathbb{R}$ be given functions, and a, b $\in \mathbb{R}$. We may define a new function.

$$af + bg : A \cap B \to \mathbb{R}, \ af + bg : x \mapsto af(x) + bg(x)$$

This is a vector field whose zero element is the zero function, $x \mapsto 0$; given a function f, its inverse is $-f = (-1)f$.

Moreover, functions form algebra. We may define the product of two functions,

$$f \cdot g : A \cap B \to \mathbb{R}, \ f \cdot g : x \mapsto f(x)g(x),$$

as well as their quotient,

$$f/g : A \cap \{z \in B : g(z) \neq 0\} \to \mathbb{R}, \ f/g : x \mapsto f(x)/g(x)$$

A third operation that combines two functions is composition. Let f: A → B and g: B → C. We define:

$$g \circ f : A \to C, \ g \circ f : x \mapsto g(f(x)).$$

For example, if f is the sine function and g is the square function, then,

$$g \circ f : \xi \mapsto \sin^2 \xi \text{ and } f \circ g : \zeta \mapsto \sin \zeta^2$$

i.e., composition is non-commutative. On the other hand, composition is associative, namely:

$$(f \circ g) \circ h = f \circ (g \circ h).$$

Note that for every function f,

$$Id \circ f = f \circ Id = f,$$

so that the identity is the neutral element with respect to function composition. This should not be confused with the fact that $x \mapsto 1$ is the neutral element with respect to function multiplication.

Example: Consider the function f that assigns the rule,

$$f : x \mapsto \frac{x + x^2 + x + \sin x^2}{x \sin x + \sin^2 x}.$$

This function can be written as:

$$f = \frac{Id + Id \cdot Id + Id \cdot \sin \circ (Id \cdot Id)}{Id \cdot \sin + Id \cdot \sin \cdot \sin}.$$

Example: Recall the n-th power functions f_n. A function P is called a polynomial of degree n if there exist real numbers $(a_i)_{i=0}^n$, with $a_n \neq 0$, such that,

$$P = \sum_{k=0}^n a_k f_k.$$

The union over all n's of polynomials of degree n is the set of polynomials. A function is called rational if it is the ratio of two polynomials.

Graph of Functions

If f is a function with domain A, then the graph of f is the set of all ordered pairs:

$$\{(x, f(x)) \mid x \in A\},$$

That is, the graph of f is the set of all points (x, y) such that $y = f(x)$. This is the same as the graph of the equation $y = f(x)$.

The graph of a function allows us to translate between algebra and pictures or geometry.

A function of the form $f(x) = mx + b$ is called a linear function because the graph of the corresponding equation $y = mx + b$ is a line. A function of the form $f(x) = c$ where c is a real number (a constant) is called a constant function since its value does not vary as x varies.

Example: Draw the graphs of the functions,

$$f(x) = 2, \ g(x) = 2x + 1.$$

Graphing functions, as you progress through calculus, your ability to picture the graph of a function will increase using sophisticated tools such as limits and derivatives. The most basic method of getting a picture of the graph of a function is to use the join-the-dots method. Basically, you pick a few values of x and calculate the corresponding values of y or $f(x)$, plot the resulting points $\{(x, f(x)\}$ and join the dots.

Example: Fill in the tables shown below for the functions,

$$f(x) = x^2, \quad g(x) = x^3, \quad h(x) = \sqrt{x}$$

and plot the corresponding points on the Cartesian plane. Join the dots to get a picture of the graph of each function.

x	$f(x) = x^2$		x	$g(x) = x^3$		x	$h(x) = \sqrt{x}$
−3			−3			0	
−2			−2			1	
−1			−1			4	
0			0			9	
1			1			16	
2			2			25	
3			3			36	

Graph of $f(x) = 1/x$:

x	$f(x) = 1/x$		x	$f(x) = 1/x$
−100			−1/1	
−10			−1/100	
−1			−1/1000	
0			0	
1			1/1000	
10			1/100	
100			1/10	

Domain and Range on Graph

The domain of the function f is the set of all values of x for which f is defined and this corresponds to all of the x-values on the graph in the xy-plane. The range of the function f is the set of all values $f(x)$ which corresponds to the y values on the graph in the xy-plane.

Example: Use the graph shown below to find the domain and range of the function,

$$f(x) = 3\sqrt{1-4x^2}.$$

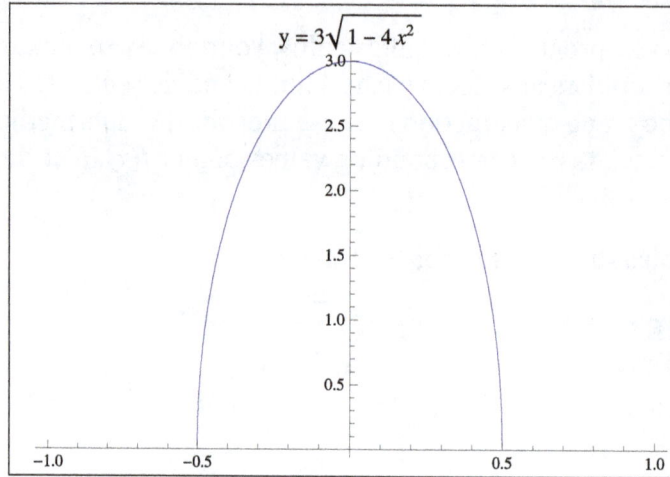

Graphing Piecewise Defined Functions

Recall that a piecewise defined function is typically defined by different formulas on different parts of its domain. The graph therefore consists of separate pieces as in the example shown below:

$$k(x) = \begin{cases} x^2 & -3 < x < 3 \\ x & 3 \le x < 5 \\ 0 & x = 5 \\ x & 5 < x \le 7 \\ \dfrac{1}{x-10} & x > 7 \end{cases}$$

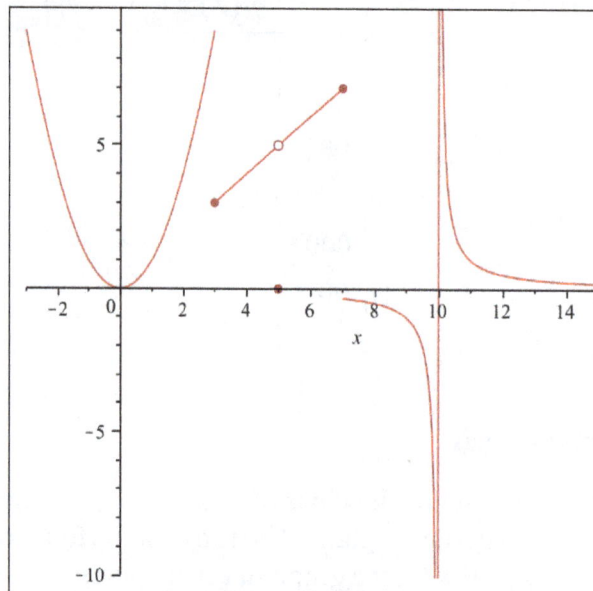

- We use a solid point at the end of a piece to emphasize that that point is on the graph. For example, the point $(3, 3)$ is on the graph whereas the point above it, $(3, 9)$, at the end of the portion of the graph of $y = x^2$ is not.

- We use a circle to denote that a point is excluded. For example the value of this function at 5 is 0, therefore the point $(5, 0)$ is on the graph as indicated with the solid dot. The point above it on the line $y = x$, $(5, 5)$, is not on the graph and is excluded from the graph. We indicate this with a circle at the point $(5, 5)$.

- Note that the formula $y = \dfrac{1}{x-10}$ does not make sense when $x = 10$. Therefore $x = 10$ is not in the domain of this function. As the values of x get closer and closer to 10 from above, the values of $\dfrac{1}{x-10}$ get larger and larger. Therefore the y values on the graph approach $+\infty$ as we approach $x = 10$ from the right. On the other hand the y values on the graph approach $-\infty$ as x approaches 10 from the left. Although there is no point on the graph at $x = 10$, the (computer generated) graph shows a vertical line at $x = 10$. This line is called a vertical asymptote to the graph.

Example: Graph the piecewise defined function,

$$f(x) \begin{cases} x & -\infty < x \le 1 \\ 2x & 1 < x < 2 \\ 1 & x = 2 \\ x^2 & x > 2 \end{cases}$$

Example: Graph the absolute value function,

$$g(x) \begin{cases} -x & x < 0 \\ x & x \ge 0 \end{cases}$$

Graphs of Equations: Vertical Line Test

It is important in calculus to distinguish between the graph of a function and graphs of equations which are not the graphs of functions. We will develop a technique called implicit differentiation to allow us to compute derivatives at (some) points on the graphs of equations which are not graphs of functions. It is therefore important to be fully aware of the relationship between graphs of equations and graphs of functions.

Recall that the defining characteristic of a function is that for every point in the domain, we get exactly one corresponding point in the range. This translates to a geometric property of the graph of the function $y = f(x)$, namely that for each x value on the graph we have a unique corresponding y value. This in turn is equivalent to the statement that if a vertical line of the form $x = a$ cuts the graph of $y = f(x)$, it cuts it exactly once. Therefore we get a geometric property which characterizes the graphs of functions: Vertical Line Property A curve in the xy-plane is the graph of a function if and only if no vertical line intersects the curve more than once.

Recall that the graph of an equation in x and y is the set of all points (x, y) in the plane which satisfy the equation. For example the graph of the equation $x^2 + y^2 = 1$ is the unit circle (circle of radius 1 centered at the origin).

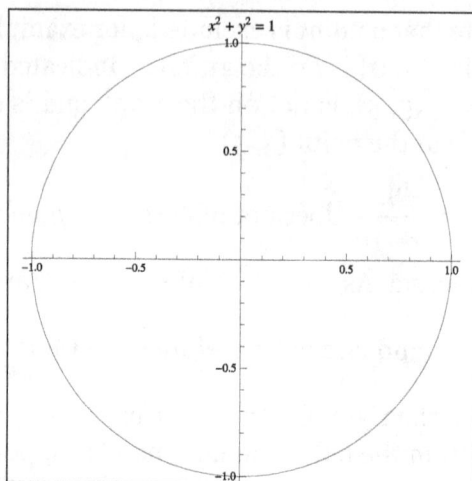

$x^2 + y^2 = 1$

If we can solve for y uniquely in terms of x in the given equation, we can rearrange the equation to look like $y = f(x)$ for some function of x. Rearranging the equation does not change the set of points which satisfy the equation, that is, it does not alter the graph of the equation. So being able to solve for y uniquely in terms of x is the algebraic equivalent of the graph of the equation being the graph of a function. This is equivalent to the graph of the equation having the vertical line property given above. Vertical Line Test The graph of an equation is the graph of a function (or equivalently if we can solve for y uniquely in terms of x) if no vertical line cuts the curve more than once.

More generally, this applies to graphs given in pieces which may be the graph of a piecewise defined function. One or several curves in the xy plane form the graph of a function (possibly piecewise defined) if no vertical line cuts the collection of curves more than once.

Example: Which of the following curves are graphs of functions?

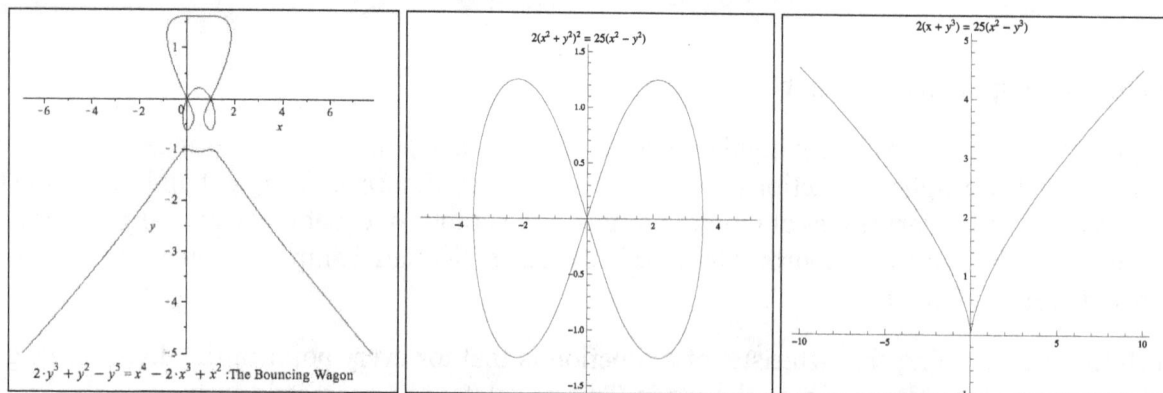

$2 \cdot y^3 + y^2 - y^5 = x^4 - 2 \cdot x^3 + x^2$:The Bouncing Wagon

$2(x^2 + y^2)^2 = 25(x^2 - y^2)$

$2(x + y^3) = 25(x^2 - y^3)$

Let us see what happens if we try to solve for y in an equation which describes a curve which not pass the vertical line tests. If we try to solve for y in terms of x in the equation:

$$x^2 + y^2 = 25$$

We get 2 new equations,

$$y = \sqrt{25 - x^2} \text{ and } y = -\sqrt{25 - x^2}.$$

The graph of the equation $x^2 + y^2 = 25$ is a circle centered at the origin (0, 0) with radius 5 and the above two equations describe the upper and lower halves of the circle respectively.

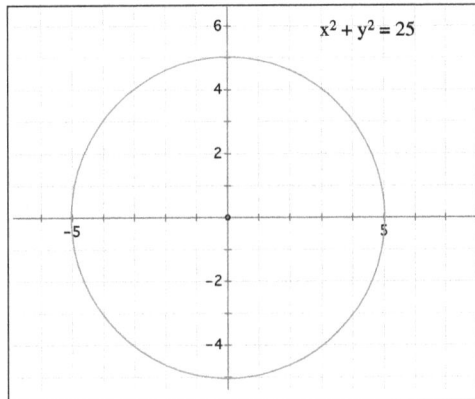

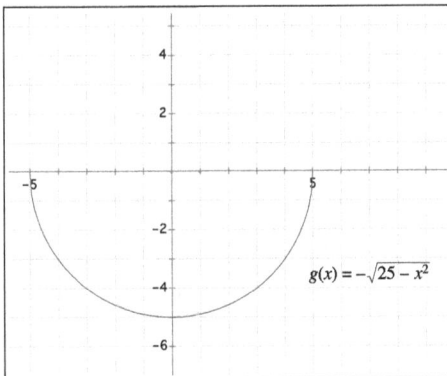
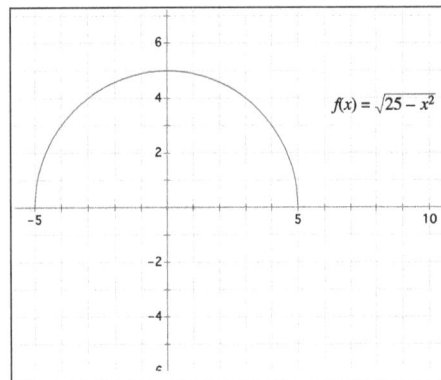

The graphs of the upper and lower halves of the circle are the graphs of functions, but the circle itself is not. Here is a catalogue of basic functions, the graphs of which you should memorize for future reference:

Lines

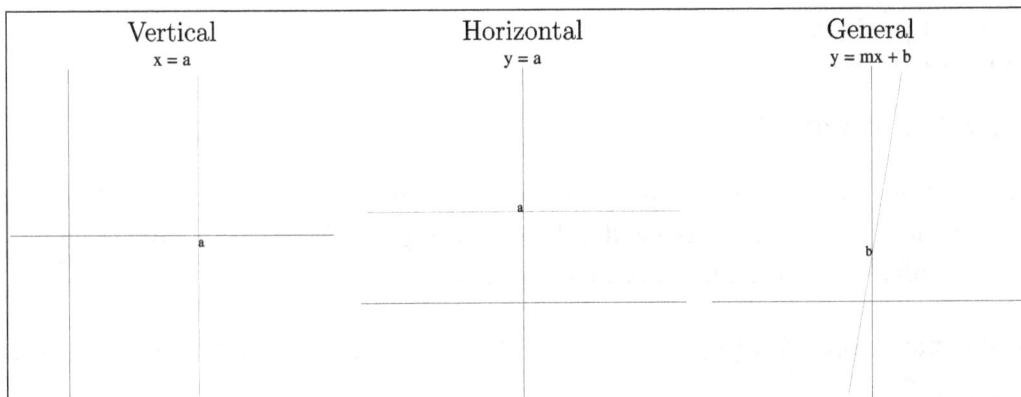

Power Functions

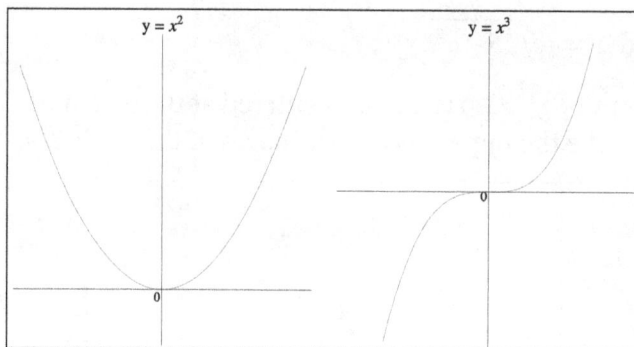

Root Functions

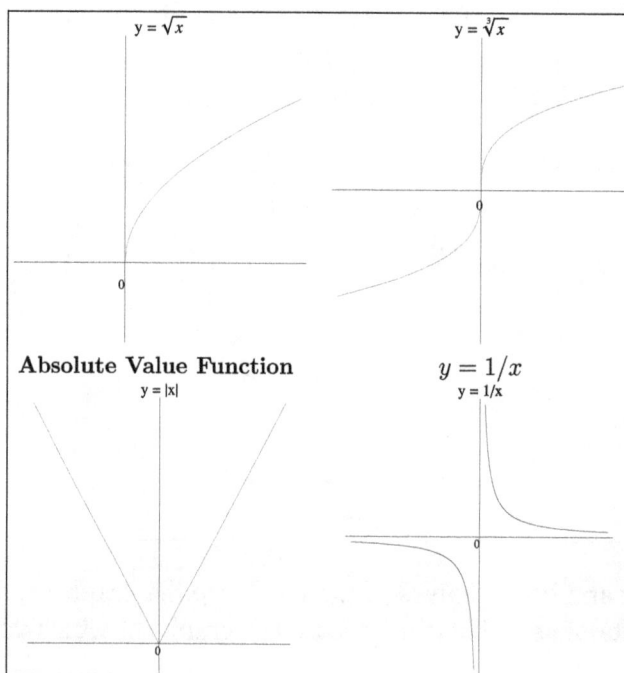

Limits and Continuity

Formal Definition of a Limit

To be able to prove results about limits and capture the concept logically, we need a formal definition of what we mean by a limit. We will only look here at the precise meaning of $\lim_{x \to \infty} f(x) = L,$ but there is a similar definition for the limit at a point.

In words, the statement $\lim_{x \to \infty} f(x) = L,$ says that $f(x)$ gets (and stays) as close as we please to L, provided we take sufficiently large x. We now try to pin down this notion of closeness.

Another way of expressing the statement above is that, if we are given any small positive number ε, then the distance between $f(x)$ and L is less than ε provided we make x large enough. We can use absolute value to measure the distance between $f(x)$ and L as $|f(x) - L|$.

How large does x have to be? Well, that depends on how small ε is.

The formal definition of $\lim_{x \to \infty} f(x) = L$, is that, given any ε > 0, there is a number M such that, if we take x to be larger than M, then the distance $|f(x) - L|$ is less than ε.

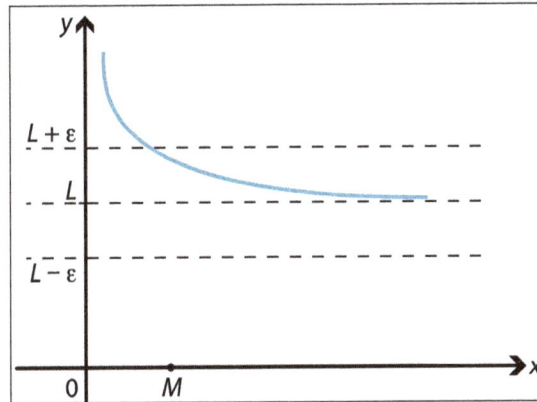

The value of $f(x)$ stays within ε of L from the point $x = M$ onwards.

For example, consider the function $f(x) = \dfrac{x+1}{x}$. We know from our basic work on limits that $\lim_{x \to \infty} f(x) = 1$. For $x > 0$, the distance is,

$$\left| f(x) - 1 \right| = \left| \frac{x+1}{x} - 1 \right| = \frac{1}{x}.$$

So, given any positive real number ε, we need to find a real number M such that, if $x > M$, then $\dfrac{1}{x} < \varepsilon$ For $x > 0$, this inequality can be rearranged to give $x > \dfrac{1}{\varepsilon}$. Hence we can choose M to be $\dfrac{1}{\varepsilon}$.

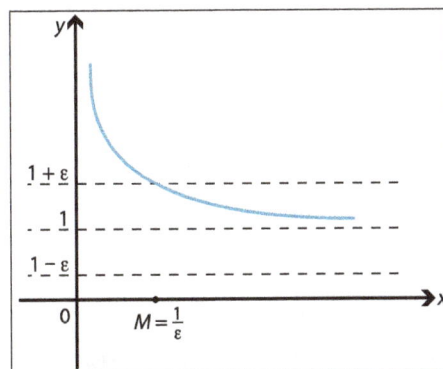

While the formal definition can be difficult to apply in some instances, it does give a very precise framework in which mathematicians can properly analyse limits and be certain about what they are doing.

Pinching Theorem

One very useful argument used to find limits is called the pinching theorem. It essentially says that if we can 'pinch' our limit between two other limits which have a common value, then this common value is the value of our limit.

Thus, if we have:

$$g\left(x\right) \le f\left(x\right) \le h(x), \text{for all } x,$$

and $\lim\limits_{x \to a} g\left(x\right) = \lim\limits_{x \to a} h\left(x\right) = L$, then $\lim\limits_{x \to a} f\left(x\right) = L.$

Here is a simple example of this,

To find $\lim\limits_{n \to \infty} \dfrac{n!}{n^n}$, we can write:

$$\frac{n!}{n^n} = \frac{n}{n} \times \frac{n-1}{n} \times \frac{n-2}{n} \times \ldots \ldots \frac{3}{n} \times \frac{2}{n} \times \frac{1}{n}$$

$$\le 1 \times 1 \times 1 \times \ldots \ldots \times 1 \times 1 \times \frac{1}{n} = \frac{1}{n},$$

where we replaced every fraction by 1 except the last. Thus we have $0 \le \dfrac{n!}{n^n} \le \dfrac{1}{n}$. Since $\lim\limits_{n \to \infty} \dfrac{1}{n} = 0,$ we can conclude using the pinching theorem that $\lim\limits_{n \to \infty} \dfrac{n!}{n^n} = 0,$

In particular, the very important limit,

$$\frac{Sin\ x}{x} \to 1 \ \ as\ x \to 0$$

(Where x is expressed in radians).

Finding Limits using Areas

One beautiful extension of the pinching theorem is to bound a limit using areas. We begin by looking at the area under the curve $y = \dfrac{1}{x}$ from $x = 1$ to $x = 1 + \dfrac{1}{n}$.

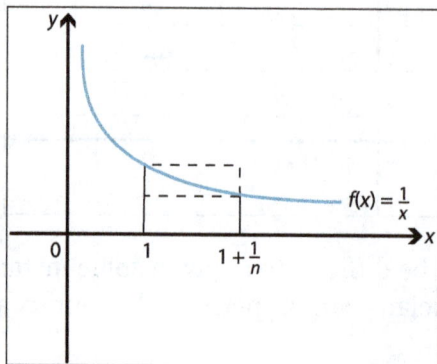

The area under the curve is bounded above and below by areas of rectangles, so we have:

$$\frac{1}{n} \times \frac{1}{1+\frac{1}{n}} \le \int_{1}^{1+\frac{1}{n}} \frac{1}{x} dx \le \frac{1}{n} \times 1.$$

Hence,

$$\frac{1}{1+n} \le \log_e \left(1+\frac{1}{n}\right) \le \frac{1}{n}.$$

Now multiplying by n, we have:

$$\frac{n}{1+n} \le n \log_e \left(1+\frac{1}{n}\right) \le 1.$$

Hence, if we take limits as $n \to \infty$, we conclude by the pinching theorem that,

$$n \log_e \left(1+\frac{1}{n}\right) \to 1 \Rightarrow \log_e \left(1+\frac{1}{n}\right)^n \to 1$$

$$\Rightarrow \left(1+\frac{1}{n}\right)^n \to e$$

That is,

$$\lim_{n \to \infty} \left(1+\frac{1}{n}\right)^n = e.$$

Limit of a Sequence

Consider the sequence whose terms begin:

$$1, \frac{1}{2}, \frac{1}{3}, \frac{1}{4}, \ldots$$

And whose general term is $\frac{1}{n}$. As we take more and more terms, each term is getting smaller in size. Indeed, we can make the terms as small as we like, provided we go far enough along the sequence. Thus, although no term in the sequence is 0, the terms can be made as close as we like to 0 by going far enough.

We say that the limit of the sequence $\left(\frac{1}{n} : n = 1, 2, 3, \ldots\right)$ is 0 and we write:

$$\lim_{n \to \infty} \frac{1}{n} = 0.$$

It is important to emphasise that we are not putting n equal to ∞ in the sequence, since infinity is not a number — it should be thought of as a convenient idea. The statement above says that the terms in the sequence $\dfrac{1}{n}$ get as close to 0 as we please (and continue to be close to 0), by allowing n to be large enough.

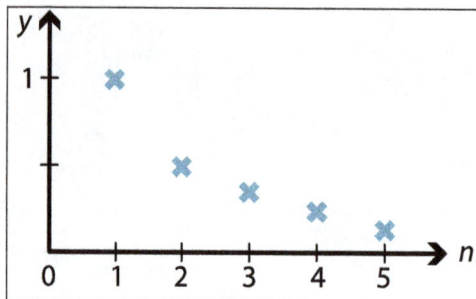

Graph of the sequence $\dfrac{1}{n}$.

In a similar spirit, it is true that we can write,

$$\lim_{n \to \infty} \frac{1}{n^a} = 0,$$

For any positive real number a. We can use this, and some algebra, to find more complicated limits.

Example: Find,

$$\lim_{n \to \infty} \frac{3n^2 + 2n + 1}{n^2 - 2}.$$

Solution: Intuitively, we can argue that, if n is very large, then the largest term (sometimes called the dominant term) in the numerator is $3n^2$, while the dominant term in the denominator is n^2. Thus, ignoring the other terms for the moment, for very large n the expression $\dfrac{3n^2 + 2n + 1}{n^2 - 2}$ is close to 3.

The best method of writing this algebraically is to divide by the highest power of n in the denominator:

$$\lim_{x \to \infty} \frac{3n^2 + 2n + 1}{n^2 - 2} = \lim_{n \to \infty} \frac{3 + \dfrac{2}{n} + \dfrac{1}{n^2}}{1 - \dfrac{2}{n^2}}.$$

Now, as n becomes as large as we like, the terms $\dfrac{2}{n}, \dfrac{1}{n^2}$ and $\dfrac{2}{n^2}$ approach 0, so we can complete the calculation and write,

$$\lim_{n \to \infty} \frac{3n^2 + 2n + 1}{n^2 - 2} = \lim_{n \to \infty} \frac{3 + \dfrac{2}{n} + \dfrac{1}{n^2}}{1 - \dfrac{2}{n^2}}$$

$$= \frac{\lim_{n \to \infty} \left(3 + \frac{2}{n} + \frac{1}{n^2} \right)}{\lim_{n \to \infty} \left(1 - \frac{2}{n^2} \right)}$$

$$= \frac{3}{1} = 3.$$

Limiting Sums

A full study of infinite series is beyond the scope of the secondary school curriculum. But one infinite series, which was studied in antiquity, is of particular importance.

Suppose we take a unit length and divide it into two equal pieces. Now repeat the process on the second of the two pieces, and continue in this way as long as you like.

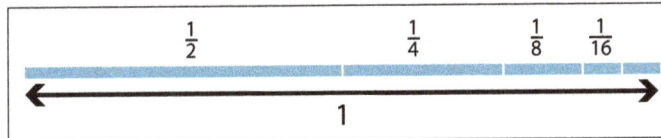

Dividing a unit length into smaller and smaller pieces.

This generates the sequence,

$$\frac{1}{2}, \frac{1}{4}, \frac{1}{8}, \frac{1}{16} \ldots\ldots,$$

Intuitively, the sum of all these pieces should be 1.

After n steps, the distance from 1 is $\frac{1}{2n}$. This can be written as:

$$\frac{1}{2} + \frac{1}{4} + \frac{1}{8} + \ldots + \frac{1}{2^n} = 1 - \frac{1}{2^n}.$$

The value of the sum approaches 1 as n becomes larger and larger. We can write this as,

$$\frac{1}{2} + \frac{1}{4} + \frac{1}{8} + \ldots + \frac{1}{2^n} \to 1 \quad \text{as } n \to \infty$$

We also write this as,

$$\frac{1}{2} + \frac{1}{4} + \frac{1}{8} + \ldots = 1.$$

This is an example of an infinite geometric series.

A series is simply the sum of the terms in a sequence. A geometric sequence is one in which each term is a constant multiple of the previous one, and the sum of such a sequence is called a geometric series. In the example considered above, each term is 1/2 times the previous term.

A typical geometric sequence has the form,

$$a, ar, ar^2, ar^3, ..., ar^{n-1}$$

where $r \neq 0$. Here a is the first term, r is the constant multiplier (often called the common ratio) and n is the number of terms.

$$Sn = a + ar + ar^2 + ar^3 + \cdots + ar^{n-1}.$$

We can easily find a simple formula for S_n. First multiply equation $a, ar, ar^2, ar^3, ..., ar^{n-1}$ by r to obtain,

$$r S_n = ar + ar^2 + ar^3 + \cdots + ar^n.$$

Subtracting equation $r S_n = ar + ar^2 + ar^3 + \cdots + ar^n.$

From equation $Sn = a + ar + ar^2 + ar^3 + \cdots + ar^{n-1}.$

Gives,

$$S_n - r S_n = a - ar^n$$

From which we have,

$$S_n = \frac{a\left(1-r^n\right)}{1-r} \quad for \ r \neq 1.$$

Now, if the common ratio r is less than 1 in magnitude, the term r^n will become very small as n becomes very large. This produces a limiting sum, sometimes written as S_∞. Thus, if $|r| < 1$,

$$S_\infty = \lim_{n \to \infty} S_n$$

$$= \lim_{n \to \infty} \frac{a\left(1-r^n\right)}{1-r} = \frac{a}{1-r}.$$

In the example considered at the start of this section, we have $a = \frac{1}{2}$ and $r = \frac{1}{2}$, hence the value of

the limiting sum is $\dfrac{\dfrac{1}{2}}{1-\dfrac{1}{2}}$ =1, as expected.

Limit of a Function at Infinity

Just as we examined the limit of a sequence, we can apply the same idea to examine the behaviour of a function $f(x)$ as x becomes very large.

For example, the following diagram shows the graph of $f(x) = \dfrac{1}{x}$, for $x > 0$. The value of the function $f(x)$ becomes very small as x becomes very large.

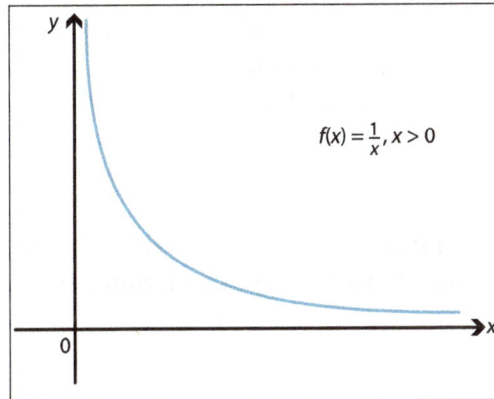

We write,

$$\lim_{x \to \infty} \frac{1}{x} = 0.$$

One of the steps involved in sketching the graph of a function is to consider the behaviour of the function for large values of x.

The following graph is of the function $f(x) = \dfrac{2x^2}{1+x^2}$. We can see that, as x becomes very large, the graph levels out and approaches, but does not reach, a height of 2.

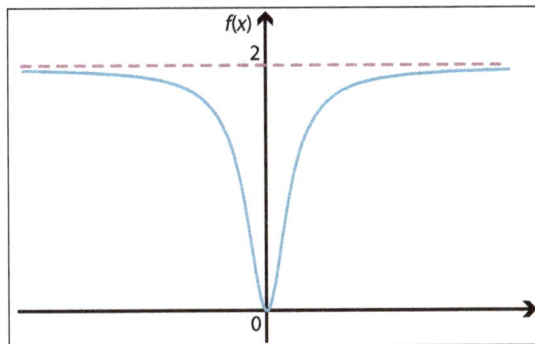

We can analyse this behaviour in terms of limits. we divide the numerator and denominator by x^2 :

$$\lim_{x \to \infty} \frac{2x^2}{1+x^2} = \lim_{x \to \infty} \frac{2}{\dfrac{1}{x^2}+1} = 2.$$

Note that as x goes to negative infinity we obtain the same limit. That is:

$$\lim_{x \to \infty} \frac{2x^2}{1+x^2} = 2.$$

This means that the function approaches, but does not reach, the value 2 as x becomes very large. The line $y = 2$ is called a horizontal asymptote for the function.

Examining the long-term behaviour of a function is a very important idea. For example, an object moving up and down under gravity on a spring, taking account of the inelasticity of the spring, is sometimes referred to as damped simple harmonic motion. The displacement, $x(t)$, of the object from the centre of motion at time t can be shown to have the form,

$$x(t) = Ae^{-\alpha t} \sin \beta t,$$

where A, α and β are positive constants. The factor $Ae^{-\alpha t}$ gives the amplitude of the motion. As t increases, this factor $Ae^{-\alpha t}$ diminishes, as we would expect. Since the factor $\sin \beta t$ remains bounded, we can write,

$$\lim_{t \to \infty} x(t) = \lim_{t \to \infty} Ae^{-\alpha t} \sin \beta t = 0.$$

In the long term, the object returns to its original position.

Limit at a Point

As well as looking at the values of a function for large values of x, we can also look at what is happening to a function near a particular point.

For example, as x gets close to the real number 2, the value of the function $f(x) = x^2$ gets close to 4. Hence we write:

$$\lim_{x \to \infty} x^2 = 4.$$

Sometimes we are given a function which is defined piecewise, such as:

$$f(x) = \begin{cases} x+3 & if \ x \le 2 \\ x & if \ x > 2. \end{cases}$$

The graph of this function is as follows:

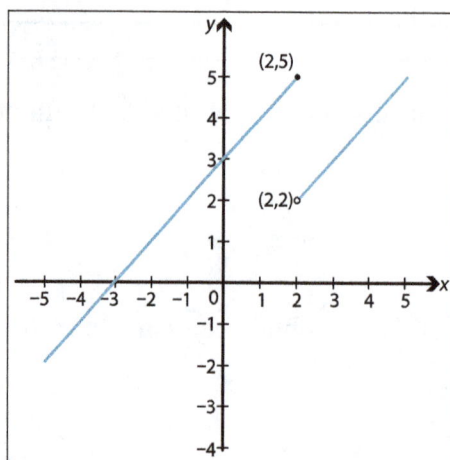

We can see from the 'jump' in the graph that the function does not have a limit at 2:

- As the x-values get closer to 2 from the left, the y-values approach 5.

- But as the x-values get closer to 2 from the right, the y-values do not approach the same number 5 (instead they approach 2).

In this case, we say that, $\lim_{x \to 2} f(x)$ does not exist.

Sometimes we are asked to analyse the limit of a function at a point which is not in the domain of the function. For example, the value $x = 3$ is not part of the domain of the function $f(x) = \dfrac{x^2 - 9}{x - 3}$.

However, if $x \neq 3$, we can simplify the function by using the difference of two squares and cancelling the (non-zero) factor $x - 3$:

$$f(x) = \frac{x^2 - 9}{x - 3} = \frac{(x-3)(x+2)}{x - 3} = x + 3, \quad \text{for } x \neq 3.$$

Now, when x is near the value 3, the value of $f(x)$ is near $3 + 3 = 6$. Hence, near the x-value 3, the function takes values near 6. We can write this as:

$$\lim_{x \to 3} \frac{x^2 - 9}{x - 3} = 6.$$

The graph of the function $f(x) = \dfrac{x^2 - 9}{x - 3}$ is a straight line with a hole at the point (3, 6).

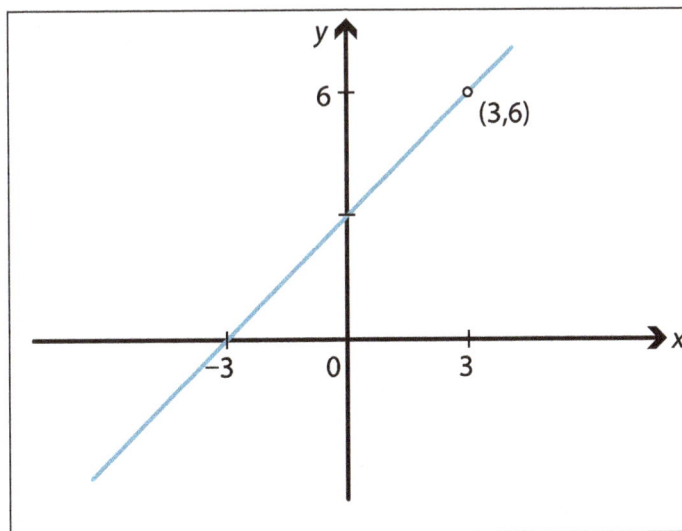

Example: Find,

$$\lim_{x \to 2} \frac{x^2 - 3x + 2}{x^2 - 4}.$$

Solution: We cannot substitute $x = 2$, as this produces 0 in the denominator. We therefore factorise and cancel the factor $x - 2$:

$$\lim_{x \to 2} \frac{x^2 - 3x + 2}{x^2 - 4} = \lim_{x \to 2} \frac{(x-2)(x-1)}{(x-2)(x+2)}$$

$$= \lim_{x \to 2} \frac{x-1}{x+2} = \frac{1}{4}.$$

Even where the limit of a function at a point does not exist, we may be able to obtain useful information regarding the behaviour of the function near that point, which can assist us in drawing its graph.

For example, the function:

$$f(x) = \frac{2}{x-1}$$

It is not defined at the point $x = 1$. As x takes values close to, but greater than 1, the values of $f(x)$ are very large and positive, while if x takes values close to, but less than 1, the values of $f(x)$ are very large and negative. We can write this as:

$$\frac{2}{x-1} \to \infty \text{ as } x \to 1^+ \quad \text{and} \quad \frac{2}{x-1} \to -\infty \text{ as } x \to 1^-.$$

The notation $x \to 1^+$ means that 'x approaches 1 from above' and $x \to 1^-$ means 'x approaches 1 from below'.

Thus, the function $f(x) = \frac{2}{x-1}$ has a vertical asymptote at $x = 1$, and the limit as $x \to 1$ does not exist. The following diagram shows the graph of the function $f(x)$. The line $y = 0$ is a horizontal asymptote.

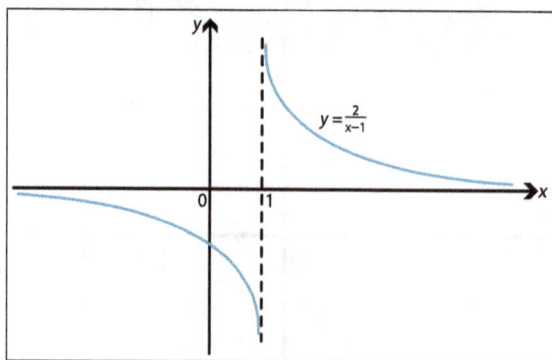

Further Examples

There are some examples of limits that require some 'tricks'. For example, consider the limit:

$$\lim_{x \to 0} \frac{\sqrt{x^2 + 4} - 2}{x^2}.$$

We cannot substitute $x = 0$, since then the denominator will be 0. To find this limit, we need to rationalise the numerator:

$$\lim_{x \to 0} \frac{\sqrt{x^2+4}-2}{x^2} = \lim_{x \to 0} \left(\frac{\sqrt{x^2+4}-2}{x^2} \times \frac{\sqrt{x^2+4}+2}{\sqrt{x^2+4}+2} \right)$$

$$= \lim_{x \to 0} \frac{(x^2+4)-4}{x^2 \left(\sqrt{x^2+4}+2 \right)}$$

$$= \lim_{x \to 0} \frac{1}{\sqrt{x^2+4}+2} = \frac{1}{4}.$$

We have implicitly assumed the following facts — none of which we can prove without a more formal definition of limit.

Limits of Algebra

Suppose that $f(x)$ and $g(x)$ are functions and that a and k are real numbers. If both $\lim_{x \to a} f(x)$ and $\lim_{x \to a} g(x)$ exist, then,

a) $\lim_{x \to a} (f(x)+g(x)) = \lim_{x \to a} f(x) + \lim_{x \to a} g(x)$

b) $\lim_{x \to a} k f(x) = k \lim_{x \to a} f(x)$

c) $\lim_{x \to a} f(x) g(x) = \left(\lim_{x \to a} f(x) \right) \left(\lim_{x \to a} g(x) \right)$

d) $\lim_{x \to a} \dfrac{f(x)}{g(x)} = \dfrac{\lim_{x \to a} f(x)}{\lim_{x \to a} g(x)}$, provided $\lim_{x \to a} g(x)$ is not equal to 0.

Continuity

When first showing students the graph of $y = x^2$, we generally calculate the squares of a number of x-values and plot the ordered pairs (x, y) to get the basic shape of the curve. We then 'join the dots' to produce a connected curve.

We can justify this either by plotting intermediary points to show that our plot is reasonable or by using technology to plot the graph. That we can 'join the dots' is really the consequence of the mathematical notion of continuity.

A formal definition of continuity is not usually covered in secondary school mathematics. For most students, a sufficient understanding of continuity will simply be that they can draw the graph of a continuous function without taking their pen off the page. So, in particular, for a function to be continuous at a point a, it must be defined at that point.

Almost all of the functions encountered in secondary school are continuous everywhere, unless they have a good reason not to be. For example, the function $f(x) = \dfrac{1}{x}$ is continuous everywhere, except at the point $x = 0$, where the function is not defined. A point at which a given function is not continuous is called a discontinuity of that function.

Here are more examples of functions that are continuous everywhere they are defined:

- Polynomials for instance, $3x^2 + 2x - 1$,

- The trigonometric functions $\sin x$, $\cos x$ and $\tan x$,

- The exponential function a^x and logarithmic function $\log_b x$ (for any bases a > 0 and b > 1).

Starting from two such functions, we can build a more complicated function by either adding, subtracting, multiplying, dividing or composing them: the new function will also be continuous everywhere it is defined.

Example: Where is the function $f(x) = \dfrac{1}{x^2 - 16}$ continuous?

Solution: The function $f(x) = \dfrac{1}{x^2 - 16}$ is a quotient of two polynomials. So this function is continuous everywhere, except at the points $x = 4$ and $x = -4$, where it is not defined.

Continuity of Piecewise-defined Functions

Since functions are often used to model real-world phenomena, sometimes a function may arise which consists of two separate pieces joined together. Questions of continuity can arise in these case at the point where the two functions are joined. For example, consider the function:

$$f(x) \begin{cases} \dfrac{x^2 - 9}{x - 3} & \text{if } x \neq 3 \\ 6 & \text{if } x = 3. \end{cases}$$

This function is continuous everywhere, except possibly at $x = 3$. We can see whether or not this function is continuous at $x = 3$ by looking at the limit as x approaches 3.

We can write:

$$\lim_{x \to 3} \frac{x^2 - 9}{x - 3} = \lim_{x \to \infty} \frac{(x - 3)(x + 3)}{x - 3} = 6.$$

Since 6 is also the value of the function at $x = 3$, we see that this function is continuous. Indeed, this function is identical with the function $f(x) = x + 3$, for all x.

Now consider the function:

$$g(x) \begin{cases} \dfrac{x^2 - 9}{x - 3} & \text{if } x \neq 3 \\ 7 & \text{if } x = 3. \end{cases}$$

The value of the function at $x = 3$ is different from the limit of the function as we approach 3, and hence this function is not continuous at $x = 3$. We can see the discontinuity at $x = 3$ in the following graph of g (x).

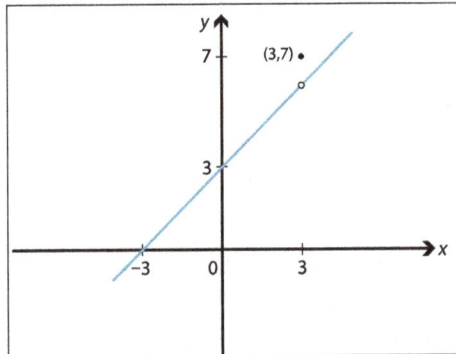

We can thus give a slightly more precise definition of a function $f(x)$ being continuous at a point a. We can say that $f(x)$ is continuous at $x =$ a if,

- $f(a)$ is defined, and
- $\lim_{x \to a} f(x) = f(a)$.

Example: Examine whether or not the function,

$$f(x) \begin{cases} x^3 - 2x + 1 & if \ x \leq 2 \\ 3x - 2 & if \ x > 2 \end{cases}$$

It is continuous at $x = 2$.

Solution: Notice that $f(2) = 2^3 - 2 \times 2 + 1 = 5$. We need to look at the limit from the right-hand side at $x = 2$. For $x > 2$, the function is given by $3x - 2$ and so,

$$\lim_{x \to 2^+} f(x) \lim_{x \to 2^+} (3x - 2) = 4.$$

In this case, the limit from the right at $x = 2$ does not equal the function value, and so the function is not continuous at $x = 2$ (although it is continuous everywhere else).

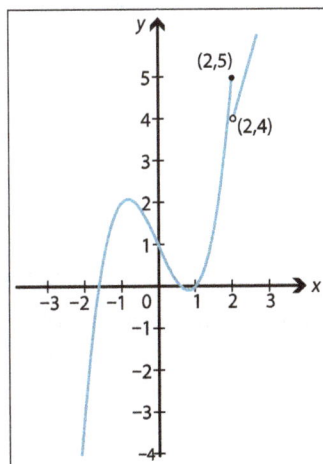

Differential Calculus

Differential Calculus is a branch of mathematics that studies derivatives and differentials of functions and their use in the investigation of functions.

The development of differential calculus into an independent mathematical discipline is associated with the names of I. Newton and G. von Leibniz (the second half of the 17th century), who formulated the basic propositions of differential calculus and clearly indicated the mutually inverse nature of the operations of differentiation and integration. Since that time differential calculus has developed with integral calculus. The (differential and integral) calculus forms the main part of mathematical analysis. The creation of the calculus launched a new era in the development of mathematics. It resulted in the appearance of a number of mathematical disciplines: the theory of series, the theory of differential equations, differential geometry, and the calculus of variations. The methods of mathematical analysis have found application in all divisions of mathematics. The number of applications of mathematics to problems of natural science and technology has expanded immeasurably. "Only differential calculus makes it possible for natural science to depict mathematically not only states, but also processes: motion".

Differential calculus is based on the following fundamental concepts of mathematics: real numbers (the number line), function, limit, and continuity. All these concepts have become crystallized and have obtained their present content in the course of the development and substantiation of the calculus. The basic idea of the differential calculus consists in studying functions locally. To be more precise, the differential calculus supplies the apparatus for studying functions whose behavior in a sufficiently small neighborhood of each point is close to the behavior of a linear function or polynomial. The central concepts of the differential calculus, the derivative and the differential, serve as such an apparatus. The concept of a derivative developed out of a large number of problems of natural science and mathematics, all of which reduced to calculating limits of the same type. The most important of these problems were the determination of the velocity of rectilinear motion of a point and the construction of a tangent to a curve. The concept of a differential is the mathematical expression of the proximity of a function to a linear function in a small neighborhood of the point being examined. In contrast to a derivative, the concept of a differential is easily carried over to mappings of one Euclidian space into another and to mappings of arbitrary normed vector spaces and is one of the principal concepts of modern nonlinear functional analysis.

Derivative

Let a function $y = f(x)$ be defined in some neighbourhood of a point x_0. Let $\Delta x \neq 0$ denote the increment of the argument and let $\Delta y = f(x_0 + \Delta x) - f(x_0)$ denote the corresponding increment of the value of the function. If there exists a (finite or infinite) limit:

$$\lim_{\Delta x \to 0} \frac{\Delta y}{\Delta x}.$$

Then this limit is said to be the derivative of the function f at x_0; it is denoted by $f'(x_0)$ $df(x_0)/dx, y', y'x, dy/dx$. Thus, by definition,

$$f'(x_0) = \lim_{\Delta x \to 0} \frac{\Delta y}{\Delta x} = \lim_{\Delta x \to 0} \frac{f(x_0 + \Delta x) - f(x_0)}{\Delta x}$$

The operation of calculating the derivative is called differentiation. If $f'(x_0)$ is finite, the function f is called differentiable at the point x_0. A function which is differentiable at each point of some interval is called differentiable in the interval.

Geometric Interpretation of the Derivative

Let C be the plane curve defined in an orthogonal coordinate system by the equation $y = f(x)$, where f is defined and is continuous in some interval J; let $M(x_0, y_0)$ be a fixed point on C, let $P(x, y)$ $(x \in J)$ be an arbitrary point of the curve C and let MP be the secant. An oriented straight line MT (T a variable point with abscissa $x_0 + \Delta x$) is called the tangent to the curve C at the point M if the angle ϕ between the secant MP and the oriented straight line tends to zero as ϕ (in other words, as the point $P \in C$ arbitrarily tends to the point M). If such a tangent exists, it is unique. Putting $x = x_0 + \Delta x, \Delta y = f(x_0 + \Delta x) - f(x_0)$ one obtains the equation $\tan \beta = \Delta y / \Delta x$ for the angle β between MP and the positive direction of the x-axis.

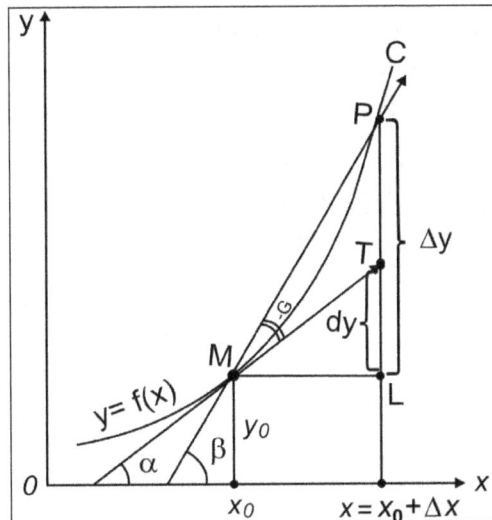

The curve C has a tangent at the point M if and only if $\lim_{\Delta x \to 0} \Delta y / \Delta x$ exists, i.e. if $f'(x_0)$ exists. The equation $\tan \alpha = f'(x_0)$ is valid for the angle α between the tangent and the positive direction of the x-axis. If $f'(x_0)$ is finite, the tangent forms an acute angle with the positive x-axis, i.e. $-\pi/2 < \alpha < \pi/2$; ; if $f'(x_0) = \infty$, the tangent forms a right angle with that axis.

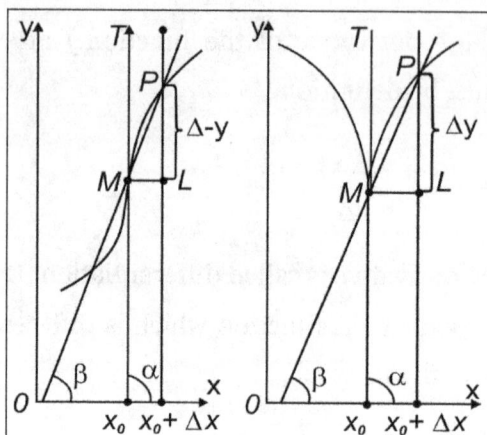

Thus, the derivative of a continuous function f at a point x_0 is identical to the slope $\tan \alpha$ of the tangent to the curve defined by the equation $y = f(x)$ at its point with abscissa x_0.

Mechanical Interpretation of the Derivative

Let a point M move in a straight line in accordance with the law $s = f(t)$. During time Δt the point M becomes displaced by $\Delta s = f(t + \Delta t) - f(t)$. The ratio $\Delta s / \Delta t$. represents the average velocity v_{av} during the time Δt. If the motion is non-uniform, v_{av} is not constant. The instantaneous velocity at the moment t is the limit of the average velocity as $\Delta t \to 0$, i.e. $v = f'(t)$ (on the assumption that this derivative in fact exists).

Thus, the concept of derivative constitutes the general solution of the problem of constructing tangents to plane curves, and of the problem of calculating the velocity of a rectilinear motion. These two problems served as the main motivation for formulating the concept of derivative.

A function which has a finite derivative at a point x_0 is continuous at this point. A continuous function need not have a finite or an infinite derivative. There exist continuous functions having no derivative at any point of their domain of definition.

The formulas given below are valid for the derivatives of the fundamental elementary functions at any point of their domain of definition (exceptions are stated):

- If $f(x) = C = \text{const}$, then $f'(x) = C' = 0$;

- If $f(x) = x$, then $f'(x) = 1$;

- $(x^\alpha)' = \alpha x^{\alpha-1}$, $\alpha = \text{const}$ $(x \neq 0, \text{if } \alpha \leq 1)$;

- $(\alpha^x)' = \alpha^x \ln \alpha$, $\alpha = \text{const} > 0$, $\alpha \neq 1$; in particular, $(e^x)' = e^x$;

- $(\log_\alpha x)' = (\log_\alpha e)/x = 1/(x \ln \alpha)$, $\alpha = \text{const} > 0$, $\alpha \neq 1$, $(\ln x)' = 1/x$;

- $(\sin x)' = \cos x$;

- $(\cos x)' = -\sin x;$
- $(\tan x)' = 1/\cos^2 x;$
- $(\cotan x)' = -1/\sin^2 x;$
- $(\arcsin x)' = 1/\sqrt{1-x^2}, x \neq \pm 1;$
- $(\arccos x)' = -1/\sqrt{1-x^2}, x \neq \pm 1;$
- $(\arctan x)' = 1/(1+x^2);$
- $(\arccotan x)' = -1/(1+x^2);$
- $(\sinh x)' = \cosh x;$
- $(\cosh x)' = \sinh x;$
- $(\tanh x)' = 1/\cosh^2 x;$
- $(\cotanh x)' = -1/\sinh^2 x.$

The following laws of differentiation are valid:

If two functions u and v are differentiable at a point x_0 then the functions,

$$cu \ (\text{where } c = \text{const}), \ u \pm v, \ uv, \ \frac{u}{v}(v \neq 0)$$

They are also differentiable at that point, and

$$(cu)' = cu',$$
$$(u \pm v)' = u' \pm v',$$
$$(uv)' = u'v + uv',$$
$$\left(\frac{u}{v}\right)' = \frac{u'v - uv'}{v^2}.$$

Theorem on the derivative of a composite function: If the function $y = f(u)$ is differentiable at a point u_0, while the function $\phi(x)$ is differentiable at a point x_0, and if $u_0 = \phi(x_0)$, then the composite function $y = f(\phi(x))$ is differentiable at x_0, and $y'_x = f'(u_0)\phi'(x_0)$ or, using another notation, $dy/dx = (dy/du)(du/dx)$.

Theorem on the derivative of the inverse function: If $y = f(x)$ and $x = g(y)$ are two mutually inverse increasing (or decreasing) functions, defined on certain intervals, and if $f'(x_0) \neq 0$ exists (i.e. is not infinite), then at the point $y_0 = f(x_0)$ the derivative $g'(y_0) = 1/f'(x_0)$ exists, or, in a different notation, $dx/dy = 1/(dy/dx)$. This theorem may be extended: If the other conditions hold and if also $f'(x_0) = 0$ or $f'(x_0) = \infty$, then, respectively, $g'(y_0) = \infty$ or $g'(y_0) = 0$.

One-sided Derivatives

If at a point x_0 the limit,

$$\lim_{\Delta x \downarrow 0} \frac{\Delta y}{\Delta x}$$

exists, it is called the right-hand derivative of the function $y = f(x)$ at x_0 (in such a case the function need not be defined everywhere in a certain neighbourhood of the point x_0; this require-ment may then be restricted to $x \geq x_0$). The left-hand derivative is defined in the same way, as:

$$\lim_{\Delta x \uparrow 0} \frac{\Delta y}{\Delta x}.$$

A function f has a derivative at a point x_0 if and only if equal right-hand and left-hand derivatives exist at that point. If the function is continuous, the existence of a right-hand (left-hand) derivative at a point is equivalent to the existence, at the corresponding point of its graph, of a right (left) one-sided semi-tangent with slope equal to the value of this one-sided derivative. Points at which the semi-tangents do not form a straight line are called angular points or cusps.

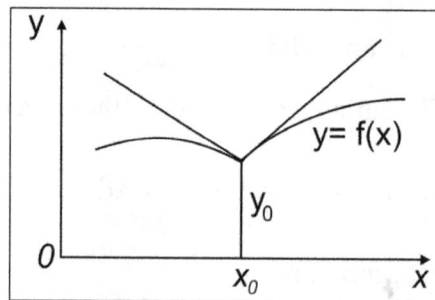

Derivatives of Higher Orders

Let a function $y = f(x)$ have a finite derivative $y' = f'(x)$ at all points of some interval; this derivative is also known as the first derivative, or the derivative of the first order, which, being a function of x, may in its turn have a derivative $y'' = f''(x)$, known as the second derivative, or the derivative of the second order, of the function f, etc. In general, the n-th derivative, or the de-rivative of order n, is defined by induction by the equation $y^{(n)} = (y^{(n-1)})'$, on the assumption that $y^{(n-1)}$ is defined on some interval. The notations employed along with $y^{(n)}$ are $f^{(n)}$, $d^n f(x) / dx^n$, and, if $n = 2, 3$, also y'', $f''(x)$, y''', $f'''(x)$.

The second derivative has a mechanical interpretation: It is the acceleration $w = d^2 s / dt^2 = f''(t)$ of a point in rectilinear motion according to the law $s = f(t)$.

Differential

Let a function $y = f(x)$ be defined in some neighbourhood of a point x and let there exist a number A such that the increment Δy may be represented as $\Delta y = A \Delta x + \omega$ with $\omega / \Delta x \to 0$ as $\Delta x \to 0$.

The term $A\Delta x$ in this sum is denoted by the symbol dy or df and is named the differential of the function $f(x)$ (with respect to the variable x) at x. The differential is the principal linear part of increment of the function (its geometrical expression is the segment LT in figure, where MT is the tangent to $y = f(x)$ at the point (x_0, y_0) under consideration).

The function $y = f(x)$ has a differential at x if and only if it has a finite derivative,

$$f'(x) = \lim_{\Delta x \to 0} \frac{\Delta y}{\Delta x} = A,$$

at this point. A function for which a differential exists is called differentiable at the point in question. Thus, the differentiability of a function implies the existence of both the differential and the finite derivative, and $dy = df(x) = f'(x)\Delta x$. For the independent variable x one puts $dx = \Delta x$, and one may accordingly write $dy = f'(x)dx$, i.e. the derivative is equal to the ratio of the differentials:

$$f'(x) = \frac{dy}{dx}.$$

The formulas and the rules for computing derivatives lead to corresponding formulas and rules for calculating differentials. In particular, the theorem on the differential of a composite function is valid: If a function $y = f(u)$ is differentiable at a point u_0, while a function $\phi(x)$ is differentiable at a point x_0 and $u_0 = \phi(x_0)$, then the composite function $y = f(\phi(x))$ is differentiable at the point x_0 and $dy = f'(u_0)\, du$, where $du = \phi'(x_0)\, dx$. The differential of a composite function has exactly the form it would have if the variable u were an independent variable. This property is known the invariance of the form of the differential. However, if u is an independent variable, $du = \Delta u$ is an arbitrary increment, but if u is a function, du is the differential of this function which, in general, is not identical with its increment.

Differentials of Higher Orders

The differential dy is also known as the first differential, or differential of the first order. Let $y = f(x)$ have a differential $dy = f'(x)dx$ at each point of some interval. Here $dx = \Delta x$ is some number independent of x and one may say, therefore, that $dx = \text{const}$. The differential dy is a function of x alone, and may in turn have a differential, known as the second differential, or the differential of the second order, of f, etc. In general, the n-th differential, or the differential of order n, is defined by induction by the equality $d^n y = d(d^{n-1}y)$, on the assumption that the differential $d^{n-1}y$ is defined on some interval and that the value of dx is identical at all steps. The invariance condition for $d^2 y, d^3 y, \ldots$, is generally not satisfied (with the exception $y = f(u)$ where u is a linear function). The repeated differential of dy has the form,

$$\delta(dy) = f''(x)dx\delta x$$

And the value of $\delta(dy)$ for $dx = \delta x$ is the second differential.

Principal Theorems and Applications of Differential Calculus

The fundamental theorems of differential calculus for functions of a single variable are usually considered to include the Rolle theorem, the Legendre theorem (on finite variation), the Cauchy theorem, and the Taylor formula. These theorems underlie the most important applications of differential calculus to the study of properties of functions — such as increasing and decreasing functions, convex and concave graphs, finding the extrema, points of inflection, and the asymptotes of a graph (Extremum; Point of inflection; Asymptote). Differential calculus makes it possible to compute the limits of a function in many cases when this is not feasible by the simplest limit theorems (indefinite limits and expressions, evaluations of). Differential calculus is extensively applied in many fields of mathematics, in particular in geometry.

Differential Calculus of Functions in Several Variables

Let a function $z = f(x, y)$ be given in a certain neighbourhood of a point (x_0, y_0) and let the value $y = y_0$ be fixed. $f(x, y_0)$ will then be a function of x alone. If it has a derivative with respect to x at x_0, this derivative is called the partial derivative of f with respect to x at (x_0, y_0); it is denoted by $f_x'(x_0, y_0)$, $\partial f(x_0, y_0)/\partial x$, $\partial f/\partial x$, z_x', $\partial z/\partial x$, or $f_x(x_0, y_0)$. Thus, by definition,

$$f_x'(x_0, y_0) = \lim_{\Delta x \to 0} \frac{\Delta_x z}{\Delta x} = \lim_{\Delta x \to 0} \frac{f(x_0 + \Delta x, y_0) - f(x_0, y_0)}{\Delta x},$$

where $\Delta_x z = f(x_0 + \Delta x, y_0) - f(x_0, y_0)$ is the partial increment of the function with respect to x (in the general case, $\partial z/\partial x$ must not be regarded as a fraction; $\partial/\partial x$ is the symbol of an operation).

The partial derivative with respect to y is defined in a similar manner:

$$f_y'(x_0, y_0) = \lim_{\Delta y \to 0} \frac{\Delta_y z}{\Delta y} = \lim_{\Delta y \to 0} \frac{f(x_0, y_0 + \Delta y) - f(x_0, y_0)}{\Delta y},$$

where $\Delta_y z$ is the partial increment of the function with respect to $\Delta_y z$. Other notations include $\partial f(x_0, y_0)/\partial y$, $\partial f/\partial y$, z_y', $\partial z/\partial y$, and $f_y(x_0, y_0)$. Partial derivatives are calculated according to the rules of differentiation of functions of a single variable (in computing z_x' one assumes $y = $ const while if z_y' is calculated, one assumes $x = $ const).

The partial differentials of $z = f(x, y)$ at (x_0, y_0) are, respectively.

$$d_x z = f_x'(x_0, y_0)dx; \quad d_y z = f_y'(x_0, y_0)dy,$$

where, as in the case of a single variable, $dx = \Delta x$, $dy = \Delta y$ denote the increments of the independent variables.

The first partial derivatives $\partial z/\partial x = f_x'(x, y)$ and $\partial z/\partial y = f_y'(x, y)$, or the partial derivatives of the first order, are functions of x and y, and may in their turn have partial derivatives with respect to x and y. These are named, with respect to the function $z = f(x, y)$, the partial derivatives of the second order, or second partial derivatives.

It is assumed that,

$$\frac{\partial}{\partial x}\left(\frac{\partial z}{\partial x}\right) = \frac{\partial^2 z}{\partial x^2}, \quad \frac{\partial}{\partial y}\left(\frac{\partial z}{\partial x}\right) = \frac{\partial^2 z}{\partial x \partial y},$$

$$\frac{\partial}{\partial x}\left(\frac{\partial z}{\partial y}\right) = \frac{\partial^2 z}{\partial y \partial x}, \quad \frac{\partial}{\partial y}\left(\frac{\partial z}{\partial x}\right) = \frac{\partial^2 z}{\partial y^2}.$$

The following notations are also used instead of $\partial^2 z / \partial x^2$:

$$z''_{xx}, \; z''_{x^2} \; \frac{\partial^2 f(x,y)}{\partial x^2}, \frac{\partial^2 f}{\partial x^2}, \; f''_{xx}(x,y), \; f''_{x^2}(x,y), \; f_{xx}(x,y);$$

and instead of $\partial^2 z / \partial x \partial y$:

$$z''_{xy}, \; \frac{\partial^2 f(x,y)}{\partial x \partial y}, \; \frac{\partial^2 f}{\partial x \partial y}, \quad f''_{xy}(x,y), \; f_{xy}(x,y),$$

One can introduce in the same manner partial derivatives of the third and higher orders, together with the respective notations: $\partial^n z / \partial x^n$ means that the function z is to be differentiated n times with respect to x; $\partial^n z / \partial x^p \partial y^q$ where $n = p + q$ means that the function z is differentiated p times with respect to x and q times with respect to y. The partial derivatives of second and higher orders obtained by differentiation with respect to different variables are known as mixed partial derivatives.

To each partial derivative corresponds some partial differential, obtained by its multiplication by the differentials of the independent variables taken to the powers equal to the number of differentiations with respect to the respective variable. In this way one obtains the n-th partial differentials, or the partial differentials of order n:

$$\frac{\partial^n z}{\partial x^n} \, dx^n, \; \frac{\partial^n z}{\partial x^p \partial y^q} \, dx^p \, dy^q.$$

The following important theorem on derivatives is valid: If, in a certain neighbourhood of a point (x_0, y_0), a function $z = f(x,y)$ has mixed partial derivatives $f''_{xy}(x,y)$ and $f''_{yx}(x,y)$, and if these derivatives are continuous at the point (x_0, y_0), then they coincide at this point.

A function $z = f(x,y)$ is called differentiable at a point (x_0, y_0) with respect to both variables x and   if it is defined in some neighbourhood of this point, and if its total increment:

$$\Delta z = f(x_0 + \Delta x, y_0 + \Delta y) - f(x_0, y_0)$$

may be represented in the form,

$$\Delta z = A\Delta x + B\Delta y + \omega,$$

where A and B are certain numbers and $\omega/\rho \to 0$ for $\rho = \sqrt{(\Delta x)^2 + (\Delta y)^2} \to 0$ (provided that the point $(x_0 + \Delta x, y_0 + \Delta y)$ lies in this neighbourhood). In this context, the expression:

$$dz = df(x_0, y_0) = A\,\Delta x + B\,\Delta y.$$

It is called the total differential (of the first order) of f at (x_0, y_0); this is the principal linear part of increment. A function which is differentiable at a point is continuous at that point (the converse proposition is not always true). Moreover, differentiability entails the existence of finite partial derivatives:

$$f_x'(x_0, y_0) = \lim_{\Delta x \to 0} \frac{\Delta_x z}{\Delta x} = A, \quad f_y'(x_0, y_0) = \lim_{\Delta y \to 0} \frac{\Delta_y z}{\Delta y} = B.$$

Thus, for a function which is differentiable at (x_0, y_0),

$$dz = df(x_0, y_0) = f_x'(x_0, y_0)\,\Delta x + f_y'(x_0, y_0)\Delta y,$$

or

$$dz = df(x_0, y_0) = f_x'(x_0, y_0)\,dx + f_y'(x_0, y_0)dy$$

If, as in the case of a single variable, one puts, for the independent variables, $dx = \Delta x$, $dy = \Delta y$.

The existence of finite partial derivatives does not, in the general case, entail differentiability (unlike in the case of functions in a single variable). The following is a sufficient criterion of the differentiability of a function in two variables: If, in a certain neighbourhood of a point (x_0, y_0), a function f has finite partial derivatives f_x' and f_y' which are continuous at (x_0, y_0), then f is differentiable at this point. Geometrically, the total differential $df(x_0, y_0)$ is the increment of the applicate of the tangent plane to the surface $z = f(x, y)$ at the point (x_0, y_0, z_0), where $z_0 = f(x_0, y_0)$.

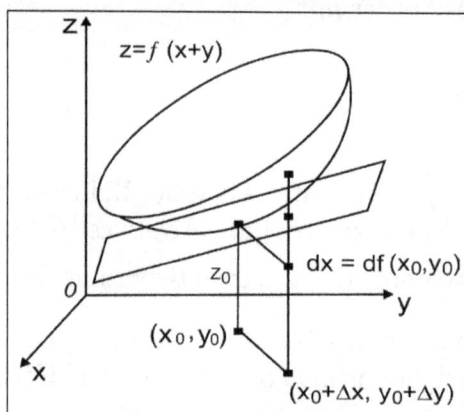

Total differentials of higher orders are, as in the case of functions of one variable, introduced by induction, by the equation,

$$d^n z = d(d^{n-1} z),$$

on the assumption that the differential $d^{n-1}z$ is defined in some neighbourhood of the point under consideration, and that equal increments of the arguments dx, dy are taken at all steps. Repeated differentials are defined in a similar manner.

Derivatives and Differentials of Composite Functions

Let $w = f(u_1, \ldots, u_m)$ be a function in m variables which is differentiable at each point of an open domain D of the m-dimensional Euclidean space R^m, and let m functions $u_1 = \phi(x_1, \ldots, x_n), \ldots, u_m = \phi_m(x_1, \ldots, x_n)$ in n variables be defined in an open domain G of the n-dimensional Euclidean space R^n. Finally, let the point $(u_1, \ldots, u_m)$, corresponding to a point $(x_1, \ldots, x_n) \in G$, be contained in D. The following theorems then hold:

A) If the functions $\phi_1, \ldots, \phi_m$ have finite partial derivatives with respect to $x_1, \ldots, x_n$, the composite function $w = f(u_1, \ldots, u_m)$ in $x_1, \ldots, x_n$ also has finite partial derivatives with respect to $x_1, \ldots, x_n$, and

$$\frac{\partial w}{\partial x_1} = \frac{\partial f}{\partial u_1} \frac{\partial u_1}{\partial x_1} + \ldots + \frac{\partial f}{\partial u_n} \frac{\partial u_n}{\partial x_1},$$

$$\frac{\partial w}{\partial x_n} = \frac{\partial f}{\partial u_n} \frac{\partial u_1}{\partial x_n} + \ldots + \frac{\partial f}{\partial u_n} \frac{\partial u_n}{\partial x_n}.$$

B) If the functions $\phi_1, \ldots, \phi_m$ are differentiable with respect to all variables at a point $(x_1, \ldots, x_n) \in G$, then the composite function $w = f(u_1, \ldots, u_m)$ is also differentiable at that point, and

$$dw = \frac{\partial f}{\partial u_1} du_1 + \ldots + \frac{\partial f}{\partial u_n} du_n,$$

where $du_1, \ldots, du_m$ are the differentials of the functions $u_1, \ldots, u_m$. Thus, the property of invariance of the first differential also applies to functions in several variables. It does not usually apply to differentials of the second or higher orders.

Differential calculus is also employed in the study of the properties of functions in several variables: finding extrema, the study of functions defined by one or more implicit equations, the theory of surfaces, etc. One of the principal tools for such purposes is the Taylor formula.

The concepts of derivative and differential and their simplest properties, connected with arithmetical operations over functions and superposition of functions, including the property of invariance of the first differential, are extended, practically unchanged, to complex-valued functions in one or more variables, to real-valued and complex-valued vector functions in one or several real variables, and to complex-valued functions and vector functions in one or several complex variables. In functional analysis the ideas of the derivative and the differential are extended to functions of the points in an abstract space.

Integral Calculus

Integral Calculus is a branch of calculus that studies the properties of and methods of computing integrals and their applications. The integral calculus is closely connected with the differential calculus and together with the latter constitutes one of the fundamental parts of mathematical analysis (or the analysis of infinitesimals). Central to the integral calculus are the concepts of the definite integral and indefinite integral of a function of a single real variable.

The fundamental concepts and theory of integral and differential calculus, primarily the relationship between differentiation and integration, as well as their application to the solution of applied problems, were developed in the works of P. de Fermat, I. Newton and G. Leibniz at the end of the 17th century. Their investigations were the beginning of an intensive development of mathematical analysis. The works of L. Euler, Jacob and Johann Bernoulli and J.L. Lagrange played an essential role in its creation in the 18th century. In the 19th century, in connection with the appearance of the notion of a limit, integral calculus achieved a logically complete form (in the works of A.L. Cauchy, B. Riemann and others). The development of the theory and methods of integral calculus took place at the end of 19th century and in the 20th century simultaneously with research into measure theory, which plays an essential role in integral calculus.

By means of integral calculus it became possible to solve by a unified method many theoretical and applied problems, both new ones which earlier had not been amenable to solution, and old ones that had previously required special artificial techniques. The basic notions of integral calculus are two closely related notions of the integral, namely the indefinite and the definite integral.

The indefinite integral of a given real-valued function on an interval on the real axis is defined as the collection of all its primitives on that interval, that is, functions whose derivatives are the given function. The indefinite integral of a function f is denoted by $\int f(x)dx$. If F is some primitive of f then any other primitive of it has the form $F + C$, where C is an arbitrary constant; one therefore writes:

$$\int f(x)dx = F(x) + C.$$

The operation of finding an indefinite integral is called integration. Integration is the operation inverse to that of differentiation,

$$\int f'(x)dx = F(x) + C, \quad d\int f(x)dx = f(x)\, dx$$

The operation of integration is linear: If on some interval the indefinite integrals exist, then for any real numbers λ_1 and λ_2, the following integral exists on this interval,

$$\int f_1(x)dx \text{ and } \int f_2(x)dx$$

$$\int [\lambda_1 f_1(x) + \lambda_2 f_2(x)]dx$$

and equals:

$$\lambda_1 \int f_1(x)dx + \lambda_2 \int f_2(x)dx.$$

For indefinite integrals, the formula of integration by parts holds: If two functions u and v are differentiable on some interval and if the integral $\int v\,du$ exists, then so does the integral $\int u\,du$, and the following formula holds,

$$\int u\,dv = uv - \int v\,du$$

The formula for change of variables holds: If for two functions f and ϕ defined on certain intervals, the composite function $f \circ \phi$ makes sense and the function ϕ is differentiable, then the integral:

$$\int f[\phi(t)]\,\phi'(t)dt.$$

Exists and equals:

$$\int f(x)dx.$$

A function that is continuous on some bounded interval has a primitive on it and hence an indefinite integral exists for it. The problem of actually finding the indefinite integral of a specified function is complicated by the fact that the indefinite integral of an elementary function is not an elementary function, in general. Many classes of functions are known for which it proves possible to express their indefinite integrals in terms of elementary functions.

The simplest examples of these are integrals that are obtained from a table of derivatives of the basic elementary functions:

1) $\int x^\alpha\,dx = \dfrac{x^{\alpha+1}}{\alpha+1} + C, \alpha \neq -1;$

2) $\int \dfrac{dx}{x} = \ln|x| + C;$

3) $\int a^x\,dx = \dfrac{a^x}{\ln\alpha} + C,\ \alpha > 0,\ \alpha \neq 1;$ in particular, $\int e^x\,dx = e^x + C;$

4) $\int \sin x\,dx = -\cos x + C;$

5) $\int \cos x\,dx = \sin x + C;$

6) $\int \dfrac{dx}{\cos^2 x} = \tan x + C;$

7) $\int \dfrac{dx}{\sin^2 x} = -\cotan x + C;$

8) $\int \sinh x\,dx = \cosh x + C;$

9) $\int \cosh x\,dx = \sinh x + C;$

10) $\int \dfrac{dx}{\cosh^2 x} = \tanh x + C;$

11) $\int \dfrac{dx}{\sinh^2 x} = -\coth x + C;$

12) $\int \dfrac{dx}{x^2 + \alpha^2} = \dfrac{1}{\alpha} \arctan \dfrac{x}{\alpha} + C = -\dfrac{1}{\alpha} \operatorname{arccotan} \dfrac{x}{\alpha} + C';$

13) $\int \dfrac{dx}{x^2 - \alpha^2} = \dfrac{1}{2\alpha} \ln \left| \dfrac{x-\alpha}{x+\alpha} \right| + C;$

14) $\int \dfrac{dx}{\sqrt{\alpha^2 - x^2}} = \arcsin \dfrac{x}{\alpha} + C = -\arccos \dfrac{x}{\alpha} + C', |x| < |\alpha|;$

15) $\int \dfrac{dx}{\sqrt{x^2 \pm \alpha^2}} = \ln |x + \sqrt{x^2 \pm \alpha^2}| + C$ (when $x^2 - \alpha^2$ is under the square root, it is assumed that $|x| > |\alpha|$).

If the denominator of the integrand vanishes at some point, then these formulas are valid only for those intervals inside which the denominator does not vanish.

The indefinite integral of a rational function over any interval on which the denominator does not vanish is a composition of rational functions, arctangents and natural logarithms. Finding the algebraic part of the indefinite integral of a rational function can be achieved by the Ostrogradski method. Integrals of the following types can be reduced by means of substitution and integration by parts to integration of rational functions:

$$\int R \left[x, \left(\dfrac{ax+b}{cx+b} \right)^{r_1}, ..., \left(\dfrac{ax+b}{cx+b} \right)^{r_m} \right] dx,$$

where $r_1, .., r_m$ are rational numbers; integrals of the form,

$$\int R(x, \sqrt{ax^2 + bx + c}) dx$$

Certain cases of integrals of differential binomials; integrals of the form,

$$\int R(\sin x, \cos x)\, dx, \quad \int R(\sinh x, \cosh x)\, dx$$

(where $R(y_1, ..., y_n)$ are rational functions); the integrals,

$$\int e^{\alpha x} \cos \beta x\, dx, \quad \int e^{\alpha x} \sin \beta x\, dx,$$

$$\int x^n \cos \alpha x\, dx, \quad \int x^n \sin \alpha x\, dx,$$

$$\int x^n \arcsin x\, dx, \quad \int x^n \arccos x\, dx,$$

$$\int x^n \arctan x\, dx, \quad \int x^n \operatorname{arccotan} x\, dx, \quad n = 0, 1, ...,$$

and many others. In contrast, for example, the integrals,

$$\int \dfrac{e^x}{x^n}\, dx, \quad \int \dfrac{\sin x}{x^n}\, dx, \quad \int \dfrac{\cos x}{x^n}\, dx, \quad n = 1, 2, ...,$$

It cannot be expressed in terms of elementary functions.

The definite integral:

$$\int_{\alpha}^{b} f(x)\, dx$$

It is a function f defined on an interval $[a, b]$ is the limit of integral sums of a specific type. If this limit exists, f is said to be Cauchy, Riemann, Lebesgue, etc. integrable.

The geometrical meaning of the integral is tied up with the notion of area: If the function $f \geq 0$ is continuous on the interval $[a, b]$, then the value of the integral,

$$\int_{\alpha}^{b} f(x)dx$$

It is equal to the area of the curvilinear trapezium formed by the graph of the function, that is, the set whose boundary consists of the graph of f, the segment $[a, b]$ and the two segments on the lines $x = a$ and $x = b$ making the figure closed, which may degenerate to points.

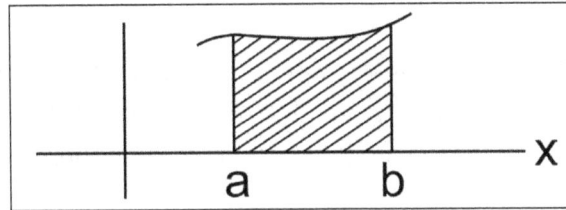

The calculation of many quantities encountered in practice reduces to the problem of calculating the limit of integral sums; in other words, finding a definite integral; for example, areas of figures and surfaces, volumes of bodies, work done by force, the coordinates of the centre of gravity, the values of the moments of inertia of various bodies, etc.

The definite integral is linear: If two functions f_1 and f_2 are integrable on an interval, then for any real numbers λ_1 and λ_2 the function:

$$\lambda_1 f_1 + \lambda_2 f_2,$$

is also integrable on this interval and

$$\int_{\alpha}^{b} \left[\lambda_1 f_1(x) + \lambda_2 f_2(x) \right] dx = \lambda_1 \int_{\alpha}^{b} f_1(x)dx + \lambda_2 \int_{\alpha}^{b} f_2(x)\, dx.$$

Integration of a function over an interval has the property of monotonicity: If the function f is integrable on the interval $[a, b]$ and if $[c, d] \subset [a, b]$, then f is integrable on $[c, d]$ as well. The integral is also additive with respect to the intervals over which the integration is carried out: If $a < c < b$ and the function f is integrable on the intervals $[a, c]$ and $[c, d]$, then it is integrable on $[a, b]$, and

$$\int\limits_{\alpha}^{b} f(x)\, dx = \int\limits_{\alpha}^{c} f(x)\, dx + \int\limits_{c}^{b} f(x)\, dx.$$

If f and g are Riemann integrable, then their product is also Riemann integrable. If $f \geq g$ on $[a, b]$, then:

$$\int\limits_{\alpha}^{b} f(x)\, dx \geq \int\limits_{\alpha}^{b} g(x)\, dx.$$

If f is integrable on $[a, b]$, then the absolute value $|f|$ is also integrable on $[a, b]$ if $-\infty < a < b\, \infty$, and

$$\left| \int\limits_{\alpha}^{b} f(x)\, dx \right| \leq \int\limits_{\alpha}^{b} |f(x)|\, dx.$$

By definition one sets:

$$\int\limits_{\alpha}^{\alpha} f(x)\,dx = 0 \quad \text{and} \quad \int\limits_{b}^{\alpha} f(x)\,dx = -\int\limits_{\alpha}^{b} f(x)\,dx, \quad a < b$$

A mean-value theorem holds for integrals. For example, if f and g are Riemann integrable on an interval $[a, b]$, if $m \leq f(x) \leq M$, $x \in [a, b]$, and if g does not change sign on $[a, b]$, that is, it is either non-negative or non-positive throughout this interval, then there exists a number $m \leq \mu \leq M$ for which:

$$\int\limits_{\alpha}^{b} f(x)g(x)\, dx = \mu \int\limits_{\alpha}^{b} g(x)\, dx.$$

Under the additional hypothesis that f is continuous on $[a, b]$, there exists in (a, b) a point ξ for which:

$$\int\limits_{\alpha}^{b} f(x)g(x)\, dx = f(\xi) \int\limits_{\alpha}^{b} g(x)\, dx.$$

In particular, if $g(x) \equiv 1$, then:

$$\int\limits_{\alpha}^{b} f(x)\,dx = f(\xi)(b - \alpha).$$

Integrals with a Variable Upper Limit

If a function f is Riemann integrable on an interval $[a, b]$, then the function F defined by,

$$F(x) = \int\limits_{\alpha}^{x} f(t)\, dt, \quad a \leq x \leq b,$$

is continuous on this interval. If, in addition, f is continuous at a point x_0, then F is differentiable at this point and $F'(x_0) = f(x_0)$. In other words, at the points of continuity of a function the following formula holds:

$$\frac{d}{dx}\int_\alpha^x f(t)\,dt = f(x).$$

Consequently, this formula holds for every Riemann-integrable function on an interval $[a, b]$, except perhaps at a set of points having Lebesgue measure zero, since if a function is Riemann integrable on some interval, then its set of points of discontinuity has measure zero. Thus, if the function f is continuous on $[a, b]$, then the function F defined by,

$$F(x) = \int_\alpha^x f(t)\,dt$$

is a primitive of f on this interval. This theorem shows that the operation of differentiation is inverse to that of taking the definite integral with a variable upper limit, and in this way a relationship is established between definite and indefinite integrals:

$$\int f(x)\,dx = \int_\alpha^x f(t)\,dt + C.$$

The geometric meaning of this relationship is that the problem of finding the tangent to a curve and the calculation of the area of plane figures are inverse operations in the above sense.

The following Newton–Leibniz formula holds for any primitive F of an integrable function f on an interval $[a, b]$:

$$\int_\alpha^b f(x)\,dx = F(b) - F(a).$$

It shows that the definite integral of a continuous function over some interval is equal to the difference of the values at the end points of this interval of any primitive of it. This formula is sometimes taken as the definition of the definite integral. Then it is proved that the integral $\int_\alpha^b f(x)\,dx$ introduced in this way is equal to the limit of the corresponding integral sums.

For definite integrals, the formulas for change of variables and integration by parts hold. Suppose, for example, that the function f is continuous on the interval (a, b) and that ϕ is continuous together with its derivative ϕ' on the interval (α, β), where (α, β) is mapped by ϕ into (a, b): $a < \phi(t) < b$ for $\alpha < t < \beta$, so that the composite $f \circ \phi$ is meaningful in (α, β). Then, for $\alpha_0, \beta_0 \in (\alpha, \beta)$, the following formulas for change of variables holds:

$$\int_{\phi(\alpha_0)}^{\phi(\beta_0)} f(x)\,dx = \int_{\alpha_0}^{\beta_0} f[\phi(t)]\,\phi'(t)\,dt.$$

The formula for integration by parts is,

$$\int_a^b u(x)\,dv(x) = u(x)v(x)\,\big|_{x=a}^{x=b} - \int_a^b v(x)\,du(x),$$

where the functions u and v have Riemann-integrable derivatives on $[\alpha, b]$.

The Newton–Leibniz formula reduces the calculation of an indefinite integral to finding the values of its primitive. Since the problem of finding a primitive is intrinsically a difficult one, other methods of finding definite integrals are of great importance, among which one should mention the method of residues and the method of differentiation or integration with respect to the parameter of a parameter-dependent integral. Numerical methods for the approximate computation of integrals have also been developed.

Generalizing the notion of an integral to the case of unbounded functions and to the case of an unbounded interval leads to the notion of the improper integral, which is defined by yet one more limit transition. The notions of the indefinite and the definite integral carry over to complex-valued functions. The representation of any holomorphic function of a complex variable in the form of a Cauchy integral over a contour played an important role in the development of the theory of analytic functions.

The generalization of the notion of the definite integral of a function of a single variable to the case of a function of several variables leads to the notion of a multiple integral.

For unbounded sets and unbounded functions of several variables, one is led to the notion of the improper integral, as in the one-dimensional case.

The extension of the practical applications of integral calculus necessitated the introduction of the notions of the curvilinear integral, i.e. the integral along a curve, the surface integral, i.e. the integral over a surface, and more generally, the integral over a manifold, which are reducible in some sense to a definite integral (the curvilinear integral reduces to an integral over an interval, the surface integral to an integral over a (plane) region, the integral over an n-dimensional manifold to an integral over an n-dimensional region). Integrals over manifolds, in particular curvilinear and surface integrals, play an important role in the integral calculus of functions of several variables; by this means a relationship is established between integration over a region and integration over its boundary or, in the general case, over a manifold and its boundary. This relationship is established by the Stokes formula which is a generalization of the Newton–Leibniz formula to the multi-dimensional case.

Multiple, curvilinear and surface integrals find direct application in mathematical physics, particularly in field theory. Multiple integrals and concepts related to them are widely used in the solution of specific applied problems. The theory of cubature formulas has been developed for the numerical calculation of multiple integrals.

The theory and methods of integral calculus of real- or complex-valued functions of a finite number of real or complex variables carry over to more general objects. For example, the theory of integration of functions whose values lie in a normed linear space, functions defined on topological groups, generalized functions, and functions of an infinite number of variables (integrals over trajectories). Finally, a new direction in integral calculus is related to the emergence and development of constructive mathematics.

Integral calculus is applied in many branches of mathematics (in the theory of differential and integral equations, in probability theory and mathematical statistics, in the theory of optimal processes, etc.), and in applications of it.

Vector Calculus

It is a field of mathematics concerned with multivariate real analysis of vectors in an inner product space of two or more dimensions; some results are those that involve the cross product can only be applied to three dimensions.

Vector Calculus	
$\varphi = f(x, y, z)$	Scalar Field
$\Delta\varphi = \vec{F}(x, y, z)$	Vector Field
$\nabla = \dfrac{\partial}{\partial x}\vec{i} + \dfrac{\partial}{\partial y}\vec{j} + \dfrac{\partial}{\partial z}\vec{k}$	Del
$\vec{F} = grad(\varphi) = \nabla\varphi$	Gradient
$\varphi = div(\vec{F}) = (\nabla \cdot \vec{F})$	Divergence
$Curl(\vec{F}) = \nabla \times \vec{F}$	Curl
$\nabla\varphi = \nabla^2\varphi = \nabla \cdot (\nabla\varphi)$	Laplacian
$\iiint_v (\nabla \cdot \vec{F})dV = \iint_m \vec{F} \cdot d\vec{S}$	Dibergence theorem

Vector Calculus is concerned with scalar fields, which associate a scalar to every point in space, and vector fields, which associate a vector to every point in space. For example, the temperature of a swimming pool is a scalar field: to each point we associate a scalar value of temperature. The water flow in the same pool is a vector field: to each point we associate a velocity vector. Vector fields are often used to model, for example, the speed and direction of a moving fluid throughout space, or the strength and direction of some force, such as the magnetic or gravitational force, as it changes from point to point.

Vector Algebra

A vector is a directed line segment, that is, a segment whose beginning (also called the vector's point of application) and end are indicated. The length of the directed line segment, which represents a vector, is called its length or magnitude. The length of vector a is denoted by |a|. Vectors are called collinear if they lie either on the same line or on parallel lines. Two vectors are said to be equal if they are collinear and have the same length and direction. All zero vectors are considered to be equal. Vector calculus, deals with free vectors.

An important role in vector algebra is played by linear operations on vectors: adding vectors and multiplying them by real numbers. The sum a + b of vectors a and b is the vector extending from the beginning of vector a to the end of vector b such that the beginning of vector b is joined to the end of vector a. The derivation of this rule is related to the parallelogram rule of vector addition, whose source is the experimental fact of the addition of forces (vector magnitudes) according to this rule. The construction of the sum of several vectors is clear from figure. The product α a of vector a and the number α is a vector that is collinear with vector a and has a length $|\alpha| \cdot |a|$ and a direction that coincides with the direction of a when $\alpha > 0$ and is opposite to that of a when $\alpha < 0$. Vector $-1 \cdot a$ is the inverse vector of a and is denoted by -a.

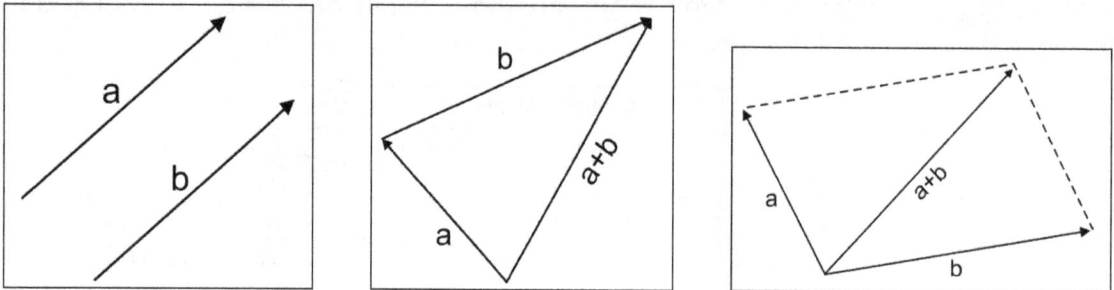

1) a + b = b + a

2) (a + b) + c = a + (b + c)

3) a + 0 = a

4) a + (-a) = 0

5) 1 · a = a

6) $\alpha(\beta a) = (\alpha\beta)a$

7) $\alpha(a + b) = \alpha a + \alpha b$

8) $(\alpha + \beta)a = \alpha a + \beta a$.

The concept of linearly dependent and linearly independent vectors is often encountered in vector algebra. Vectors a_1, a_2,.......a_n are called linearly dependent if there exist such numbers α_1, α_2,...... α_n, of which at least one of them differs from zero, that the linear combination $\alpha_1 a_1 + \cdots + \alpha_n a_n$ of these vectors is equal to zero. Vectors a_1, a_2,..............a_n that are not linearly dependent are called linearly independent. Let us note that any three nonzero vectors not lying in one plane are linearly independent.

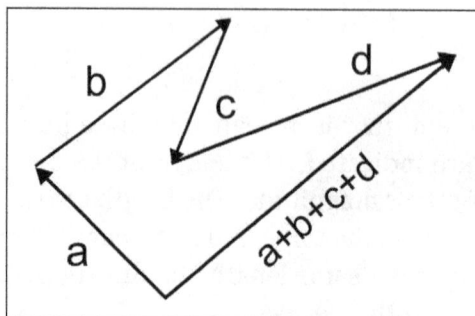

The vectors of Euclidean space have the following property: there exist three linearly independent vectors, and any arbitrary four vectors are linearly dependent. This property characterizes the three-dimensionality of the set of vectors under consideration. In conjunction with the properties listed above, the indicated property implies that the set of all vectors of Euclidean space forms a so called vector space. The linearly independent vectors e_1, e_2, and e_3 form a basis. Any vector a can be uniquely resolved in terms of basis vectors: a = X e_1 + Y e_2 + Z e_3 the coefficients X, Y, and Z are called the coordinates (components) of vector a in the given basis. If vector a has coordinates X, Y, and Z, this can be written as a = $\{X, Y, Z\}$. Three mutually orthogonal (perpendicular) vectors, whose lengths are equal to one and which are usually denoted by i, j, and k, form a socalled orthonormalized basis. If these vectors are located with their initial points at one point O, they form a rectangular Cartesian coordinate system in space. The coordinates X, Y, Z of any point M in this system are defined as the coordinates of the vector O M. The linear operations on vectors, indicated previously, correspond to analogous operations on their coordinates: if the coordinates of vectors a and b are $\{X_1, Y_1, Z_1\}$ and $X, Y_2, Z_2\}$ respectively, then the coordinates of the sum a + b of these vectors are $\{X_1 + X_2, Y_1 + Y_2, Z_1 + Z_2\}$ the coordinates of vector λ_a are $\{\lambda X_1, \lambda Y_1, \lambda Z_1\}$.

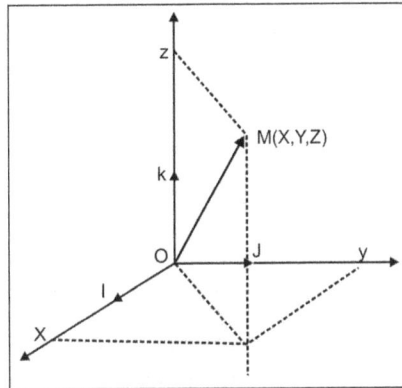

The development and application of vector algebra is closely connected with various types of products of vectors: scalar, vector, and mixed. The concept of the scalar product of vectors arises, for example, in examining the work performed by a force F along a given path S: the work is equal to $|F||S|\cos\varphi$, where φ is the angle between vectors F and S. Mathematically, the scalar product of vectors a and b is defined as the number denoted by (a, b) and equal to the product of the magnitudes of these vectors and of the cosine of the angle between them,

$(a, b) = |a|\ |bx|\ \cos\varphi$

The quantity $|b|\cos\varphi$ is called the projection of vector b on the axis determined by vector a and is denoted by $proj_a b$. Therefore, $(a, b) = |a|\ proj_a b$. In particular, if a is a unit vector ($|a| = 1$), then (a, b). The following properties of the scalar product are obvious,

- $(a, b) = (b, a)$;

- $(\lambda_a, b) = \varphi(a,b)$;

- $(a + b, c) = (a,c) + (b,c)$;

- $(a, a) \geq 0)$.

where equality with zero occurs only for a = o. If vectors a and b have the coordinates $\{X_1, Y_1, Z_1\}$ and $\{X_2, Y_2, Z_2\}$ respectively, in an orthonormalized basis i, j, k, then,

$$(a, b) = X_1 X_2 + Y_1 Y_2 + Z_1 Z_2$$

$$|a| = \sqrt{X_1^2 + Y_1^2 + Z_1^2}$$

$$\cos \varphi = \frac{X_1 X_2 + Y_1 Y_2 + Z_1 Z_2}{\sqrt{X_1^2 + Y_1^2 + Z_1^2} \sqrt{X_2^2 + Y_2^2 + Z_2^2}}$$

The definition of a vector product requires use of the concept of a left- and right-handed ordering of three vectors. The ordered triplet of vectors *a*, *b*, *c* (a is the first vector, b, the second, and c, the third), starting at the same point and not lying in one plane, is called right-handed (left-handed) if the vectors are situated in the same way as the thumb, index, and middle fingers, respectively, of the right (left) hand. Figure shows right-handed (on the right) and left-handed (on the left) triplets of vectors.

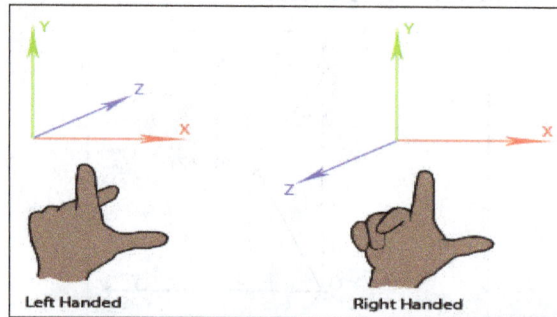

The vector product of vectors a and b is the vector denoted by [a, b] and satisfying the following requirements: (1) the length of vector [a, b] is equal to the product of the lengths of vectors a and and of the sine of the angle φ between them (thus, if a and b are collinear, then [a, b] = o); and (2) if a and b are noncollinear, then [a, b] is perpendicular to both vectors a and b and is directed so that the triplet of vectors a, b, [a, b] is right-handed. The vector product has the following properties,

- [a, b] = [b, a]

- [(λa), b] = [a, b]

- [c, (a + b)] = [c, a] + [c, b]

- [a, [b, c]] = b (a, c) - c(a, b)

- ([a, b], [c, d]) = (a, c)(b, d) - (a, d)(b, c)

If, in an orthonormalized basis i, j, k forming a right handed triplet, vectors a and b have the coordinates $\{X_1 Y_1 Z_1\}$ and $\{X_2 Y_2 Z_2\}$, respectively, then [a, b] =$\{Y_1 Z_2 - Y_2 Z_1, Z_1 X_2 - Z_2 X_1, X_1 Y_2 - X_2 Y_1\}$.

The concept of vector product is connected with various problems in mechanics and physics. For example, the velocity v of a point *M* of an object rotating around an axis/with an angular velocity of ω is [ω,r], where r = *OM*.

The mixed product of vectors a, b, and c is the scalar product of vector $[a, b]$ and vector c: ($[a, b]$, c). It is denoted by abc. The mixed product of vectors a, b, and c that are not parallel to the same plane is numerically equal to the volume of the parallelepiped formed by bringing the vectors a, b, and c to a common initial point; its sign is positive if the triplet a, b, c is right handed and negative if the Triplet is left-handed.

If vectors a, b, and c are parallel to the same plane, then abc = 0. The property that $abc = bca = cab$ also holds true. If the coordinates of vectors a, b, and c in an orthonormalized basis i, j, k, which forms a right-handed triplet, are respectively equal to $\{X_1, Y_1, Z_1\}$, $\{X_2, Y_2, Z_2\}$, and $\{X_3, Y_3, Z_3\}$, then,

$$abc = \begin{vmatrix} X_1 & Y_1 & Z_1 \\ X_2 & Y_2 & Z_2 \\ X_3 & Y_3 & Z_3 \end{vmatrix}$$

Vector functions of scalar arguments in mechanics, physics, and differential geometry frequent use is made of the concept of a vector function of one or several scalar arguments. If a definite vector r is in correspondence to every value of a variable t of a certain set {t} according to a known law, then one says that a vector function r = r(t) is specified by the set {t}. Since vector r is defined by coordinates $\{x, y, z\}$, the specification of the vector function r = r(t) is equivalent to the specification of three scalar functions: $x = x(t)$, $y = y(t)$, and $z = z(t)$. The concept of vector function becomes particularly obvious if it is converted to a so-called hodograph of this function, that is, to the locus of the ends of all vectors r(t) joined to the coordinate origin O. If, in this case, one considers the argument t to be time, then the vector function r(t) represents the law of motion of point M moving along curve L—the hodograph of r(t).

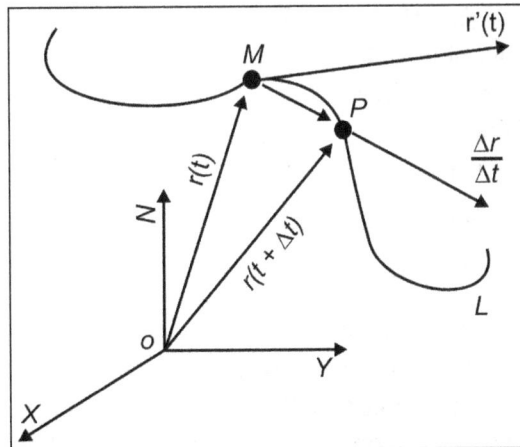

The concept of derivative plays an important role in the study of vector functions. This concept is introduced in the following way: to the argument t is added the increment $\Delta t \neq 0$ and the vector Δr = r(t + Δt) - r(t) (the increment Δr is vector Vector $\overrightarrow{MP}$ in figure) is multiplied by 1/Δt. The limit of the expression $\Delta r/\Delta t$ as $\Delta t \rightarrow 0$ is called the derivative of the vector function and is denoted by r'(t) or dr/dt. The derivative is the vector that is tangent to the hodograph L at the given point M. If the vector function is regarded as the law of motion of a point along the curve L, then the derivative r'(t) is equal to the velocity of this point's motion. The rules for computing the derivatives of various products of vector functions are similar to the rules of finding the derivatives of the products of ordinary functions. For example,

$$(r_1, r_2)' = (r_1', r_2) + (r_1, r_2')$$
$$[r_1, r_2]' = [r_1', r_2] + [r_1, r_2']$$

In differential geometry the vector functions of one argument are used for the definition of curves. Vector functions of two arguments are used for the specification of surfaces.

Vector Analysis

In mechanics, physics, and geometry the concepts of scalar and vector fields are frequently used. The temperature of a non-uniformly heated plate and the density of a nonhomogeneous body are physical examples of plane and three-dimensional scalar fields, respectively. A vector field is a set of all the velocity vectors of particles of a steady flow of fluid. Other examples of vector fields are the gravitational force field and the electrical and magnetic potentials of an electromagnetic field.

For the mathematical specification of scalar and vector fields, scalar and vector functions are used, respectively. It is clear that the density of an object is a scalar point-function and that the velocity field of the particles of steady liquid flow is a vector point-function. The mathematical apparatus of field theory is usually called vector analysis. For the geometric characterization of a scalar field one uses the concepts of contour lines and equipotential surfaces. The contour line of a plane scalar field is a line on which the function that defines the field has a constant value. The equipotential surface of a spatial field is defined in an analogous way. An example of a contour line is an isotherm—the contour line of the scalar temperature field of a no uniformly heated plate.

We now consider equipotential surfaces (lines) of a scalar field which pass through a given point M. The maximum change of the function f which defines the field at this point, occurs along a normal to this surface (line) at the point M. This change is characterized by the gradient of the scalar field. The gradient is a vector that is directed along the normal to the equipotential surface (line) at point M in the direction of the increasing f at this point. The magnitude of the gradient is equal to the derivative of f in the indicated direction. The gradient is denoted by the symbol grad f. If the basis is i, j, k, then grad f has the coordinates $\{\partial f/\partial x, \partial f/\partial y, \partial f/\partial z\}$ for a plane field the gradient coordinates are $\{\partial f/\partial x, \partial f/\partial y,\}$ the gradient of a scalar field is a vector field.

A number of concepts are introduced to characterize vector fields: vector lines, vector tubes, circulations of a vector field, and divergence and curl (rotor) of a vector field. In some region Ω, let a vector field be denoted by the vector function a(M) of a variable point M of Ω. A line L in the region Ω is called a vector line if the vector tangent at each of its points M is directed along vector a(M).

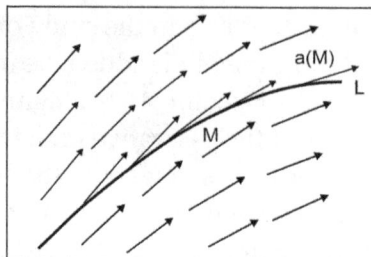

If the field a*(M)* is a velocity field of the particles of a steady flow of a fluid, then the vector lines of this field are the trajectories of the fluid particles. The part of space in Ω that consists of vector lines is called a vector tube. If one is dealing with the vector field of velocities of the particles of a steady flow of a fluid, then the vector tube is the part of space that a certain fixed volume of fluid "sweeps out" in its motion.

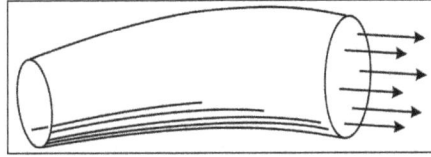

Let AB be a smooth curve in Ω, l the length of arc AB measured from point A to the variable point *M* of this line, and t the unit vector tangent to AB at *M*. The circulation of the field a*(M)* along the curve AB is the expression,

$$\int_{AB} (a, t) dt$$

If a*(M)* is a force field, then the circulation of a along AB is the work performed by this field along the path AB.

The divergence of vector field a*(M)* which has the coordinates P, Q, R in the basis i, j, k, is defined as the sum $\partial p / \partial x + \partial Q / \partial y + \partial R / \partial z$ and is denoted by the symbol div a. For example, the divergence of the gravitational field created by a certain mass distribution is equal to the density (volumetric) $\rho(x, y, z)$ of this field multiplied by 4π.

The curl (or rot) of vector field a*(M)* is a vector characterizing the "rotational component" of this field. The curl of field a is denoted by rot a or curl a. If P, Q, R are the coordinates of a in the basis i, j, k, then,

$$rot\ a = \left\{ \frac{\partial R}{\partial y} - \frac{\partial Q}{\partial z}, \frac{\partial P}{\partial z} - \frac{\partial R}{\partial x}, \frac{\partial Q}{\partial x} - \frac{\partial P}{\partial y} \right\}$$

Let field a be the velocity field of a fluid flow. We place a small wheel with vanes at a given point of the flow and orient its axis in the direction of rot a at this point. Then the flowrate will be a maximum, and its value will be ½[rot a]. The gradient of a scalar field and the divergence and curl of a vector field are usually called the fundamental differential operations of vector analysis. The following equations, relating these operations, hold true:

- grad (*fh*) = *f* grad *h* + *h* grad *f*

- div (*f*a) = (a, grad *f*) + *f* div a

- rot (*f*a) = *f* rot a + [grad *f*, a]

- div [a, b] = (b, rot a) - (a, rot b)

Vector field a*(M)* is called potential if it is the gradient of some scalar field *f(M)*. In this case, the field *f(M)* is called the potential of vector field a. In order that the field a, whose coordinates *P, Q, R* have continuous partial derivatives, be a potential field, it is necessary and sufficient that the curl of this field vanish. If a potential field is given in a simply connected region Ω, then the potential *f(M)* of this field can be found from the formula,

$$f = \int_{AM} (a, t)\, dl$$

where AM is any smooth curve connecting a fixed point A of Ω with point *M*, t is the unit vector tangent to the curve AM, and l is the length of arc AM measured from point A.

Vector field *a(M)* is called solenoidal or tubular if it is the curl of some field b*(M)*. Field b*(M)* is called the vector potential of field a. In order that a be solenoidal, it is necessary and sufficient that the divergence of this field vanish. An important role in vector analysis is played by integral relations: Ostrogradskii's formula, also designated the fundamental formula of vector analysis, and Stokes' formula. Let V be a region whose boundary T consists of a finite number of pieces of smooth surfaces and *n* be the unit vector of the exterior normal to T. Let vector field a*(M)* be given in the region V such that div a is a continuous function. Then the following holds true:

$$\iiint_v div\ adv\ =\ \iint r\ (a,n)\ d\sigma$$

This is known as Ostrogradskii's formula.

If a is the velocity field of a steady flow of incompressible fluid, then (a, n) dσ is the volume of fluid that passes through an area dσ on the boundary r in a unit of time. Therefore, the right-hand side of equation above is the flow of fluid through the boundary r of body V per unit time. Because, in the case being considered, div a characterizes the intensity of the fluid sources, Ostrogradskii's formula expresses the following obvious fact: the flow of fluid through a closed surface r is equal to the amount of fluid generated by all the sources inside r. Let a continuous and differentiable vector field which has a continuous curl rot a be assigned in a region Ω. Let r be an orientable surface consisting of a finite number of pieces of smooth surface, *n* the unit normal to r, t the unit vector tangent to the edge *y* of the surface r, and l the length of the arc *y*. The following relation, called Stokes' formula, holds true,

$$\iint_r (n, rot\ a)\, d\sigma = \oint_r (a, b)\, dl$$

Equation $\iint_r (n, rot\ a)\, d\sigma = \oint_r (a, b)\, dl$ expresses the following physical fact: the intensity of the curl of a vector field a through the surface r is equal to the circulation of this field along the curve *y*. Ostrogradskii's formula is the source of the invariant (independent of the coordinate system) definition of the fundamental operations of vector analysis. For example, from this formula, it follows that,

$$div\ a = \lim_{v \to 0} \frac{\iint_r (a, n)\, d\sigma}{V}$$

Because the expression,

$$\iint_r (a, n)\, d\sigma$$

is the flow of fluid through r and,

$$\frac{1}{V} \iint (a,n) \, d\sigma$$

is the magnitude of this flow per unit volume, the definition of div a by means of equation

$div\ a = \lim\limits_{v \to 0} \dfrac{\iint_r (a,n)\,d\sigma}{V}$ indicates that div a characterizes the flux of the source at a given point.

Applications of Vector Calculus

For a continuously differentiable function of several real variables, a point P, that is a set of values for the input variables, which is viewed as a point in R_n, which is critical if all of the partial derivatives of the function are zero at P or equivalently, if it's gradient is zero. The critical values are the values of the function at the critical points.

Vectors sounds are complicated, but they are common when giving directions. For example, telling someone to walk to the end of a street before turning left and walking five more blocks is an example of using vectors to give directions. Navigating by air and by boat is generally done using vectors. Planes are given a vector to travel, and they use their speed to determine how far they need to go before turning or landing. Flight plans are made using a series of vectors.

Sports instructions are based on using vectors. Wide receivers playing American football, for example, might run a route where they run seven yards down the field before turning left 45 degrees and running in that direction. Sports commentary also depends on vectors. Only a few sports have fields with grids, so discussions revolve around the direction and speed of the player.

Multivariable Calculus

Multivariable calculus is a branch of calculus in one variable to calculus with functions of more than one variable. In single variable calculus, we study the function of single variable whereas in multivariable calculus we study with two or more variables.

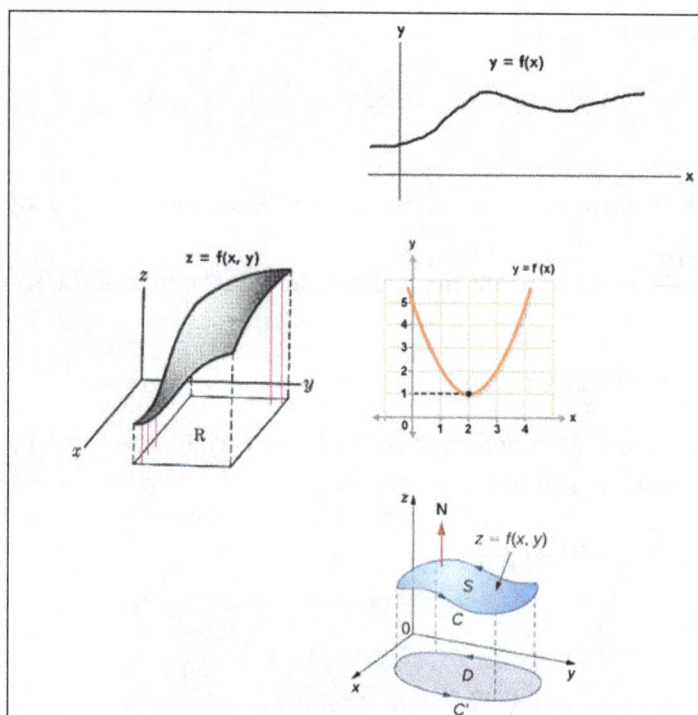

Partial Derivatives

It is derivative of a function of two or more variables with respect to one of those variables, with the other held constant. It is used in vector calculus and differential geometry.

Let's assume $F(x, y)$ to be a function with two variables. By keeping y constant and differentiable F (assuming F is differentiable) with respect to variable x. What we obtain is the partial derivative of F with respect to x and is denoted by $\partial F/\partial x$ or Fx.

In the same way, partial derivative of F with respect to y is denoted by $\partial F/\partial y$ OR Fy.

Critical Point of Function of Two Variables

Critical point of a function with two variables is a point where the partial derivative of first order are equal to 0. To find a critical point we must first take the derivative of the function. Then set that derivative equal to 0 and solve for x. Each value of x that w get is known as the critical number.

Let's find the critical point of function F defined by $F(x, y) = x^2 + y^2$. We start with finding the first order partial derivative.

- $F_x(x, y) = 2x$

- $F_y(x, y) = 2y$

Now we will solve the equation $F_x(x, y) = 0$ and $F_y(x, y) = 0$ simultaneously.

- $F_x(x, y) = 2x = 0$

- $F_y(x, y) = 2y = 0$

The solution of the above equation is the ordered pair (0, 0). The graph of $F(x, y) = x^2 + y^2$ states that at the critical point (0, 0) f has a minimum value.

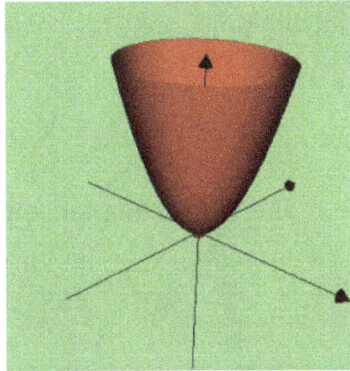

Maxima and Minima of Functions of Two Variable

After getting the critical point we do the second derivative test to determine if a critical point is a relative maximum or relative minimum. The meaning of maximum and minimum dose not state that the relative maximum or minimum is the largest/smallest value that the function will ever take. It just states that in some region around point (a, b) the function will always be smaller/larger than F(a, b). It is possible for the function to be larger/smaller outside of that region.

Firstly, we need to figure out how many second derivatives we have,

$$\frac{\partial^2 F}{\partial x^2} = F_{xx}$$

$$F_{xy} = \frac{\partial^2 F}{\partial_x \partial y}$$

$$\frac{\partial^2 F}{\partial y \partial x} = F_{yx}$$

$$\frac{\partial^2 F}{\partial y^2} = F_{yy}$$

It is interesting to note that,

$$F_{xy} = \frac{\partial^2 F}{\partial_x \partial y} = \frac{\partial^2 F}{\partial_y \partial x} = F_{yx}$$

The second derivative test states that when we have a critical point (x_o, y_o) of Function of two variables and have to calculate the partial derivative,

Let A $= F_{xx}(x_0, y_0), B = F_{xy}(x_0, y_0), C = F_{xy}(x_0 y_0)$

If AC $-$ B² > 0 and A > 0 then it is the minimum and when A < 0 is the maximum. When, AC $-$ B² < 0 then it is called the saddle point.

And when, $AC - B^2 = 0$ then we cannot conclude whether it will be the minimum, maximum or the saddle point.

Optimization problems involve optimizing functions in two variables using first and second order.

Let us look at some problems closely.

Example: $F(x, y) = x^2 + 2y^2 - x^2y$

Solution: Critical point occurs where F_x and F_y are simultaneously 0.

$F_x = 2x - 2xy = 2x(1-y)$

$F_y = 4y - x^2$

$F_x = 0$ if $x = 0$ or $y = 1$

Using this in the equation $F_y = 0$

If $x = 0$, $y = 0$

If $y = 1$ then $4 - x^2 = 0$

Therefore, we have $(0,0)$, $(2,1)$, and $(-2,1)$

Now using the second partial test to classify, $D = F_{xx} \cdot F_{yy} - (F_{xy})^2 = (2 - 2y) \cdot (4) - (-2x)^2$

- At $(0,0)$ D= 8 and $F_{xx} = 2$ therefore we have a minimum.

- At $(2,1)$ D = -16 is the saddle point.

- At $(-2,1)$ D = -16 = 0 therefore is a saddle point.

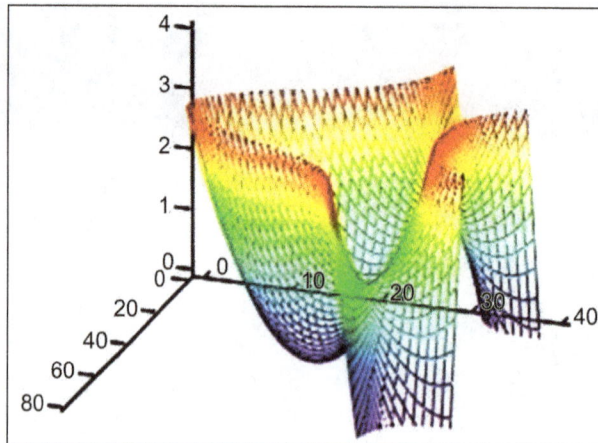

Application of Multivariable Calculus

Multivariable calculus is useful considering that most natural phenomenon are non-linear and can be best described by using multivariable calculus and differential equation. For example relationship between speed, position and acceleration can be defined by multivariable calculus and differential equation.

References

- Calculus: ipracticemath.com, Retrieved 2 March, 2019
- Graphs-of-Functions: nd.edu, Retrieved 17 July, 2019
- Differential-calculus: thefreedictionary.com, Retrieved 7 May, 2019
- Differential-calculus: encyclopediaofmath.org, Retrieved 19 August, 2019
- Integral-Calculus: thefreedictionary.com, Retrieved 21 January, 2019
- Integral-calculus: encyclopediaofmath.org, Retrieved 23 March, 2019
- Vector-calculus, physics, science: assignmentpoint.com, Retrieved 28 February, 2019
- Vector-Calculus: thefreedictionary.com, Retrieved 8 June, 2019
- Multivariable-calculus: toppr.com, Retrieved 30 April, 2019

The Derivative

The derivative of a function of a real variable is used to measure the sensitivity to change of the function value in relation to a change in its input value. Some of the concepts studied in relation to derivatives are local extrema of functions and rules for finding derivatives. This chapter closely examines these key concepts of derivatives to provide an extensive understanding of the subject.

In Calculus, derivative is the rate of change of a function with respect to a variable. Derivatives are fundamental to the solution of problems in calculus and differential equations. In general, scientists observe changing systems (dynamical systems) to obtain the rate of change of some variable of interest, incorporate this information into some differential equation, and use integration techniques to obtain a function that can be used to predict the behaviour of the original system under diverse conditions.

Geometrically, the derivative of a function can be interpreted as the slope of the graph of the function or, more precisely, as the slope of the tangent line at a point. Its calculation, in fact, derives from the slope formula for a straight line, except that a limiting process must be used for curves. The slope is often expressed as the "rise" over the "run," or, in Cartesian terms, the ratio of the change in y to the change in x. For the straight line shown in the figure, the formula for the slope is $(y_1 - y_0)/(x_1 - x_0)$. Another way to express this formula is $[f(x_0 + h) - f(x_0)]/h$, if h is used for $x_1 - x_0$ and $f(x)$ for y. This change in notation is useful for advancing from the idea of the slope of a line to the more general concept of the derivative of a function.

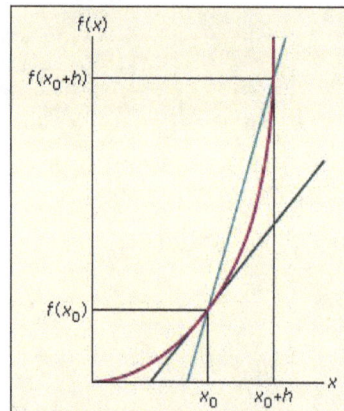

The slope, or instantaneous rate of change, for a curve at a particular point $(x_0, f(x_0))$ can be determined by observing the limit of the average rate of change as a second point $(x_0 + h, f(x_0 + h))$ approaches the original point.

For a curve, this ratio depends on where the points are chosen, reflecting the fact that curves do not have a constant slope. To find the slope at a desired point, the choice of the second point needed to calculate the ratio represents a difficulty because, in general, the ratio will represent only an average slope between the points, rather than the actual slope at either point. To get around this difficulty, a limiting process is used whereby the second point is not fixed but specified by a variable, as h in the ratio for the straight line above. Finding the limit in this case is a process of finding a number that the ratio approaches as h approaches 0, so that the limiting ratio will represent the actual slope at the given point. Some manipulations must be done on the quotient $[f(x_0 + h) - f(x_0)]/h$ so that it can be rewritten in a form in which the limit as h approaches 0 can be seen more directly. Consider, for example, the parabola given by x^2. In finding the derivative of x^2 when x is 2, the quotient is $[(2 + h)^2 - 2^2]/h$. By expanding the numerator, the quotient becomes $(4 + 4h + h^2 - 4)/h = (4h + h^2)/h$. Both numerator and denominator still approach 0, but if h is not actually zero but only very close to it, then h can be divided out, giving $4 + h$, which is easily seen to approach 4 as h approaches 0.

To sum up, the derivative of $f(x)$ at x_0, written as $f'(x_0)$, $(df/dx)(x_0)$, or $Df(x_0)$, is defined as,

$$\lim_{h \to 0} \left[f(x_0 + h) - f(x_0) \right] / h$$

if this limit exists.

Derivative of Function

The derivative of a function f at a point x, written f '(x) is given by,

$$f'(x) = \lim_{\Delta x \to 0} \frac{f(x + \Delta x) - f(x)}{\Delta x}$$

if this limit exists.

Graphically, the derivative of a function corresponds to the slope of its tangent line at one specific point. The following illustration allows us to visualise the tangent line (in blue) of a given function at two distinct points. Note that the slope of the tangent line varies from one point to the next. The value of the derivative of a function therefore depends on the point in which we decide to evaluate it.

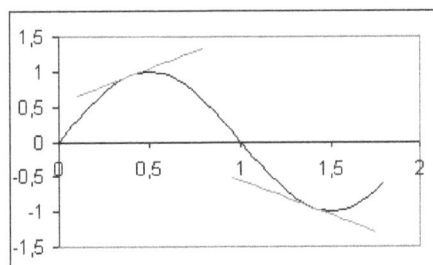

Notation

Here, we represent the derivative of a function by a prime symbol. For example, writing $f'(x)$ represents the derivative of the function f evaluated at point x. Similarly, writing $(3x + 2)'$ indicates we are carrying out the derivative of the function $3x + 2$. The prime symbol disappears as soon as the derivative has been calculated.

Derivatives of usual Functions

Below you will find a list of the most important derivatives. Although these formulas can be formally proven, we will only state them here.

Constant Function

Let $f(x) = k$, where k is some real constant.

Then,

$$f'(x) = (k)' = 0$$

Examples:

$$(8)' = 0$$

$$(-5)' = 0$$

$$(0,2321)' = 0$$

The identity function $f(x) = x$

Let $f(x) = x$, the identity function of x.

Then,

$$f'(x) = (x)' = 1$$

A function of the form x^n.

Let $f(x) = x^n$, a function of x, and n a real constant. We have,

$$f(x) = (x^n) = nx^{n-1}$$

Examples:

$$\left(x^4\right)' = 4x^{4-1} = 4x^3$$

$$\left(x^{1/2}\right)' = 1/2x^{\frac{1}{2}-1} = 1/2x^{-1/2}$$

$$\left(x^{-2}\right)' = -2x^{-2-1} = -2x^{-3}$$

$$\left(x^{-\frac{1}{3}}\right)' = \left(-\frac{1}{3}\right)x^{-\frac{1}{3}-1} = \left(-\frac{1}{3}\right)x^{-\frac{4}{3}}$$

$(x^n)' = n\,x^{n-1}$ Rule

- The rule mentioned above applies to all types of exponents (natural, whole, and fractional). It is however essential that this exponent is constant. Another rule will need to be studied for exponential functions (of type a^x).

- The identity function is a particular case of the functions of form x$_n$ (with $n = 1$) and follows the same derivation rule: $(x) = (x^1) = 1\,x^{1-1} = 1\,x^0 = 1.$

- It is often the case that a function satisfies this form but requires a bit of reformulation before proceeding to the derivative. It is the case of roots (square, cubic, etc.) representing fractional exponents.

Examples:

$$\sqrt{x} = x^{1/2} \rightarrow \left(\sqrt{x}\right)' = \left(x^{\frac{1}{2}}\right)' = \frac{1}{2}x^{\left(\frac{1}{2}-1\right)} = \frac{1}{2}x^{-1/2}$$

$$\sqrt[3]{x} = x^{1/3} \rightarrow \left(\sqrt[3]{x}\right)' = \left(x^{1/3}\right)' = \frac{1}{3}x^{\left(\frac{1}{3}-1\right)} = \frac{1}{3}x^{-2/3}$$

- Beware of rational functions. For example, the function $\dfrac{1}{x^4}$ cannot be differentiated in the same manner as the function x^4. You must first reformulate the function so that "x" is a numerator, forcing us to change its exponent's sign.

Examples:

$$\left(\frac{1}{x^4}\right)' = x^{-4} \rightarrow \left(\frac{1}{x^4}\right) = \left(x^{-4}\right)' = -4x^{-4-1} = -4x^{-5} = -\frac{4}{x^5}$$

$$\frac{1}{x^{3/2}} = x^{-3/2} \rightarrow \left(\frac{1}{x^{3/2}}\right)' = \left(x^{-\frac{3}{2}}\right)' = -\frac{3}{2}x^{-\frac{3}{2}} = -\frac{3}{2}x^{-\frac{5}{2}} = -\frac{3}{2x^{\frac{5}{2}}}$$

- Finally, a derivate can greatly be simplified by proceeding first, if possible, to an algebraic simplification.

Example:

$$\frac{x^2}{x^3\sqrt{x}} = \frac{x^2}{x^3 x^{1/2}} = x^{2-3-1/2} = x^{-3/2}$$

That is how the derivative of $\dfrac{x^2}{x^3\sqrt{x}}$ is greatly facilitated by carrying out the derivative of $x^{-3/2}$.

$$\left(\frac{x^2}{x^3\sqrt{x}}\right)' = \left(x^{-\frac{3}{2}}\right)' = -\frac{3}{2}x^{-\frac{3}{2}-1} = -\frac{3}{2}x^{-\frac{5}{2}}$$

Exponential Function

It is very easy to confuse the exponential function a^s with a function of the form x^n since both have exponents. They are, however, quite different. In an exponential function, the exponent is a variable.

Given the exponential function $f(x) = a^x$ where a >0. We have:

$$f'(x) = (a^x)' = a^x \ln(a)$$

Examples:

$$\left(3^x\right)' = 3^x \ln(3)$$

$$\left(\left(\frac{1}{2}\right)^x\right)' = \left(\frac{1}{2}\right)^x \ln\left(\frac{1}{2}\right)$$

Function e^x

Let the function $f(x) = e^x$.

Then,

$$f(x) = (e^x) = e^x$$

Here is a special case of the previous rule since the function $f(x) = e^x$ is an exponential function with a = e.

Therefore, $f'(x) = (e^x)' = e^x \ln(e) = e^x(1) = e^x$.

Logarithmic Function ln x

Given the logarithmic function $f(x) = \ln x$.

We have:

$$f'(x) = (\ln x)' = \frac{1}{x}.$$

Basic Derivation Rules

We will generally have to confront not only the functions presented above, but also combinations of these: multiples, sums, products, quotients and composite functions. We therefore need to present the rules that allow us to derive these more complex cases.

Constant Multiples

Let k be a real constant and $f(x)$ any given function.

Then,

$$(k\,f(x))' = k\,f'(x).$$

In other words, we can forget the constant which will remain unchanged and only derive the function of x.

Examples:

$$\left(4x^2\right)' = 4\left(x^2\right)' = 4\left(2x\right) = 8x$$

$$\left(-5e^x\right)' = -5\left(e^x\right) = -5e^x$$

$$\left(12\ln x\right)' = 12\left(\ln x\right)' = 12\left(\frac{1}{x}\right) = \frac{12}{x}$$

Addition and Subtraction of Functions

Let $f(x)$ and $g(x)$ be two functions.

Then,

$$(f(x) \pm g(x))' = f'(x) \pm g'(x).$$

When we derive a sum or a subtraction of two functions, the previous rule states that the functions can be individually derived without changing the operation linking them.

Example:

$$\left(e^x + x^5\right)' = \left(e^x\right)' + \left(x^5\right)' = e^x + 5x^4$$

Example:

$$\left(\ln x - \frac{1}{x^2} + 8\right)' = \left(\ln x\right)' - \left(x^{-2}\right) + \left(8\right)'$$

$$= \frac{1}{x} - \left(-2x^{-3}\right) + 0$$

$$= \frac{1}{x} + \frac{2}{x^3}$$

Example:

$$\left(3\sqrt{x} + 2x - \frac{8}{x}\right)' = \left(3x^{1/2}\right)' + \left(2x\right)' - \left(8x^{-1}\right)'$$

$$= 3\left(x^{\frac{1}{2}}\right)' + 2\left(x\right)' - 8\left(x^{-1}\right)'$$

$$= 3\left(\frac{1}{2}x^{-\frac{1}{2}}\right) + 2\left(1\right) - 8\left(-x^{-2}\right)$$

$$= \frac{3}{2}x^{-\frac{1}{2}} + 2 + 8x^{-2}$$

Product Rule

Let $f(x)$ and $g(x)$ be two functions. Then the derivate of the product:

$$(f(x)\, g(x)\,)' = f'(x)\, g(x) + f(x)\, g'(x)$$

Examples:

$$\left(x^3 e^x\right)' = \left(x^3\right)' e^x + x^3 \left(e^x\right)'$$

$$= 3x^2 e^x + x^3 e^x$$

$$\left(3\sqrt{x}\,\ln x\right)' = \left(3\sqrt{x}\right)'\ln x + 3\sqrt{x}\left(\ln x\right)'$$

$$= 3\left(x^{\frac{1}{2}}\right)\ln x + 3\sqrt{x}\left(\ln x\right)'$$

$$= 3\left(\frac{1}{2}x^{\frac{1}{2}-1}\right)\ln x + 3\sqrt{x}\,\frac{1}{x}$$

$$= \frac{3}{2}x^{-\frac{1}{2}}\ln x + 3x^{-\frac{1}{2}}$$

Quotient Rule

Let $f(x)$ and $g(x)$ be two functions. Then the derivative of the quotient:

$$\left(\frac{f(x)}{g(x)}\right)' = \frac{f'(x)g(x) - f(x)g'(x)}{\left[g(x)\right]^2}$$

Examples:

$$\left(\frac{x^3}{e^x}\right)' = \frac{\left(x^3\right)' e^x - x^3\left(e^x\right)'}{\left(ex\right)^2}$$

$$= \frac{3x^2 e^x - x^3 e^x}{\left(e^x\right)^2}$$

$$= \frac{x^2 e^x\left(3-x\right)}{\left(e^x\right)^2}$$

$$= \frac{x^2\left(3-x\right)}{e^x}$$

Examples:

$$\left(\frac{3\sqrt{x}}{\ln x}\right)' = \frac{\left(3\sqrt{z}\right)'\ln x - 3\sqrt{x}\left(Inx\right)'}{\left(Inx\right)^2} = \frac{3\left(x^{\frac{1}{2}}\right)'\ln x - 3\sqrt{x}\left(\ln x\right)'}{\left(\ln x\right)^2}$$

$$= \frac{3\left(\frac{1}{2}x^{\frac{1}{2}-1}\right)'\ln x - 3\sqrt{x}\,\frac{1}{x}}{\left(\ln x\right)^2} = \frac{3x - \frac{1}{2}\ln x - 6x - \frac{1}{2}}{2\left(\ln x\right)^2}$$

$$= \frac{3x - \frac{1}{2}\left(\ln x - 2\right)}{2\left(Inx\right)^2}$$

Derivative of Composite Functions

A composite function is a function with form $f(g(x))$. A composite function is in fact a function that contains another function. If you have a function that can be broken down into many parts, where each part is in it a function and where these parts are not linked by addition, subtraction, product or division, you usually have a composite function.

For example, the function $f(x) = e^{x^3}$ is a composite function. We can write it as $f(g(x))$ where $g(x) = x^3$.

Unlike the function $f(x) = x^3 e^x$ which is not a composite function. It is only the product of functions.

Here are a few examples of composite functions:

- $f(x) = ln(x^2 + 2x + 1)$

  We can write this function as $f(g(x)) = ln(g(x))$ where $g(x) = x^2 + 2x + 1$;
- $f(x) = e^{3x-5}$

  We can write this function as $f(g(x)) = e^{g(x)}$ where $g(x) = 3x - 5$;
- $f(x) = (ln(x) + 3x - e^x)^4$

  We can write this function as $f(g(x)) = (g(x))^4$ where,

  $g(x) = ln(x) + 3x - e^x$.

Chain Rule

Let f and g be two functions. Then the derivative of the composite function $f(g(x))$ is:

$$(f(g(x)))' = f'(g(x)) \, g'(x)$$

$$(f(u))' = f'(u) \, u', \text{ where: } u = g(x).$$

The chain rule states that when we derive a composite function, we must first derive the external function (the one which contains all others) by keeping the internal function as is and then multiplying it with the derivative of the internal function. If the latter is also composite, the process is repeated. Be alert as the internal function could also be a product, a quotient.

Chain Derivatives of usual Functions

In concrete terms, we can express the chain rule for the most important functions as follows:

If $u = g(x)$ represents any given function of x,

- $(u^n)' = n \, u^{n-1} \, u'$
- $(a^u)' = a^u ln(a) u'$
- $(e^u)' = e^u u'$
- $(ln \, u)' = \dfrac{1}{u} \times u'$

Examples:

$$\left[ln\left(x^2+2x+1\right)\right]' = \frac{1}{x^2+2x+1}\left(x^2+2x+1\right)'$$

$$= \frac{1}{x^2+2x+1}\left(2x+2\right)$$

$$= \frac{2x+2}{x^2+2x+1}$$

$$\left[\left(lnx+3x-e^x\right)^4\right]' = 4\left(ln\,x+3x-e^x\right)^3\left(ln\,x+3x-e^x\right)'$$

$$= 4\left(ln\,x+3x-e^x\right)^3\left(\frac{1}{x}+3-e^x\right)$$

$$\left(e^{3x-5}\right)' = e^{3x-5}\left(3x-5\right)'$$

$$= e^{3x-5}\cdot 3$$

Below are additional examples that demonstrate that many rules may be necessary for one derivative.

Examples:

$$\left(\left[ln\left(3x^3-9e^x\right)\right]^3\right)' = 3\left[ln\left(3x^3-9e^x\right)\right]^2\cdot\left[ln\left(3x^3-9e^x\right)\right]'$$

$$= 3\left[ln\left(3x^3-9e^x\right)\right]^2\cdot\frac{1}{3x^3-9e^x}\cdot\left(3x^3-9e^x\right)'$$

$$= 3\left[ln\left(3x^3-9e^x\right)\right]^2\cdot\frac{1}{3x^3-9e^x}\cdot\left(9e^2-9e^x\right)$$

$$\left[e^{x\ln x}\right]' = e^{x\ln x}\cdot\left(x\ln x\right)'$$

$$= e^{x\ln x}\left[\left(x\right)'In\,x+x\left(In\,x\right)'\right] \qquad \text{(Product rule)}$$

$$= e^{x\ln x}\left(1.In\,x+x.\frac{1}{x}\right)$$

$$= e^{x\ln x}\left(In\,x+1\right)$$

$$\left[\frac{x^2+1}{\left(2x+1\right)^{\frac{1}{2}}}\right]' = \frac{\left(x^2+1\right)'\cdot\left(2x+1\right)^{\frac{1}{2}}-\left(x^2+1\right)\cdot\left[\left(2x+1\right)^{\frac{1}{2}}\right]'}{\left[\left(2x+1\right)^{\frac{1}{2}}\right]^2} \quad \text{(quotient rule)}$$

$$= \frac{2x.\left(2x+1\right)^{\frac{1}{2}}-\left(x^2+1\right).\left[\left(2x+1\right)^{\frac{1}{2}}\right]'}{2x+1}$$

$$= \frac{2x.(2x+1)^{\frac{1}{2}} - (x^2+1).\frac{1}{2}(2x+1)^{-\frac{1}{2}}.(2x+1)'}{2x+1}$$

$$= \frac{2x.(2x+1)^{\frac{1}{2}} - (x^2+1).\frac{1}{2}(2x+1)^{-\frac{1}{2}}.2}{2x+1}$$

$$= \frac{2x.(2x+1)^{\frac{1}{2}} - (x^2+1).(2x+1)^{-\frac{1}{2}}}{2x+1}$$

Evaluation of the Slope of the Tangent at One Point

As we mentioned at the very beginning, the derivative function $f'(x)$ represents the slope of the tangent line at $f(x)$ at all points x. We will often have to evaluate this slope at a specific point.

To evaluate the slope of the tangent of the function $f(x)$ at the point x = 1 for example, we most certainly cannot calculate $f(1)$ and derive this value. We would then obtain a slope of 0 since $f(1)$ is a constant. Instead, we need to find the derivative $f'(x)$ at all points and then evaluate it at x = 1. We will use the notation $f'(a)$ to represent the derivative of the function f evaluated at the point x = a.

Example: Evaluate the slope of the function $f(x) = x^3\, e^s$ at the point x = 0.

We are looking to calculate $f'(0)$. We must first find the derivative at all points, $f'(x)$. Yet earlier we demonstrated that $f'(x) = (x^3\, e^s)' = 3x^2\, e^s + x^3\, e^s$.

Evaluated at x = 0, we obtain $f'(0) = 3.\, 0^2\, e^0 + 0^3\, e^0 = 0$. The slope of the function $f'(x)$ = x3esis therefore zero at x = 0. We will let you verify that this is not the case at point x = 1.

Increasing and Decreasing Functions

There is a direct relationship between the growth and decline of a function and the value of its derivative at one point:

- If the value of the derivative is negative at a given point, this indicates that the function is decreasing at that point.

- If the value of the derivative is positive at a given point, this indicates that the function is increasing at that point.

Example:

- Find the derivative of the function $f(x) = (x^2 - 4)^3$.

- What is the slope of the tangent of $f(x)$ at the point = 1?

- Is the function $f(x)$ increasing or decreasing at the point x = 1?

- Find all points where the slope of $f(x)$ is 0.

Solution:

- We need to derive the composite function u^3, where $u = x^2 - 4$. Consequently, we need to use the chain derivative.

$$f'(x) = \left[\left(x^2 - 4 \right)^3 \right]'$$

$$= 3\left(x^2 - 4 \right)^2 . \left(x^2 - 4 \right)'$$

$$= 3\left(x^2 - 4 \right)^2 . 2x$$

$$= 6x\left(x^2 - 4 \right)^2$$

- At point $x = 1$, the slope of the tangent of the function f is:

$$f'(1) = 6(1)\left(1^2 - 4\right)^2 = 6(1)(-3)^2 = 54$$

- Since the slope is positive at $x = 1$, the function $f(x)$ is increasing at this point.

- The slope is 0 at points like $f'(x) = 0$. We therefore need to find the values of x so that,

$$6x\left(x^2 - 4 \right)^2 = 0$$

$x = 0$, $x = -2$ and $x = 2$ are the values sought.

Local Extrema of Functions

Let a function $y = f(x)$ be defined in a δ-neighborhood of a point x_0, where $\delta > 0$. The function $f(x)$ is said to have a local (or relative) maximum at the point x_0, if for all points $x \neq x_0$ belonging to the neighborhood $(x_0 - \delta, x_0 + \delta)$ the following inequality holds:

$$f(x) \leq f(x_0).$$

If the strict inequality holds for all points $x \neq x_0$ in some neighborhood of x_0,

$$f(x) < f(x_0),$$

then the point x_0 is a strict local maximum point.

Similarly, we define a local (or relative) minimum of the function $f(x)$. In this case, the following inequality is valid for all points $x \neq x_0$ of the δ-neighborhood $(x_0 - \delta, x_0 + \delta)$ of the point x_0:

$$f(x) \geq f(x_0).$$

Accordingly, a strict local minimum is described by the inequality:

$$f(x) > f(x_0).$$

The concepts of local maximum and local minimum are united under the general term local extremum. The word "local" is often ommitted for brevity, so it is said simply about maxima and minima of functions.

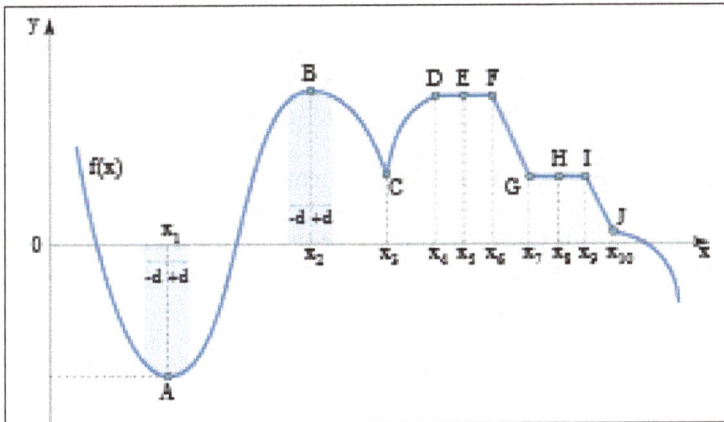

Figure schematically shows the different extrema points. The point $A(x_1)$ is a strict local minimum point, since there exists a δ-neighborhood $(x_1 - \delta, x_1 + \delta)$, in which the following inequality holds:

$$f(x) > f(x_1) \forall x \in (x_1 - \delta, x_1 + \delta).$$

Similarly, the point $B(x_2)$ is a strict local maximum point. At this point, we have the inequality:

$$f(x) > f(x_2) \forall x \in (x_2 - \delta, x_2 + \delta).$$

The number δ at each point may be different.

The subsequent points are classified as follows:

- $C(x_3)$ is a strict minimum point;

- $D(x_4)$ is a non-strict maximum point;

- $E(x_5)$ is a non-strict maximum or minimum point;

- $F(x_6)$ is a non-strict maximum point;

- $G(x_7)$ is a non-strict minimum point;

- $H(x_8)$ is a non-strict maximum or minimum point;

- $I(x_9)$ is a non-strict maximum point;

- $J(x_{10})$ – there is no extremum.

Necessary Condition for an Extremum

The points at which the derivative of the function $f(x)$ is equal to zero are called the stationary points.

The points at which the derivative of the function $f(x)$ is equal to zero or does not exist are called the critical points of the function. Consequently, the stationary points are a subset of the set of critical points.

A necessary condition for an extremum is formulated as, If the point x_0 is an extremum point of the function $f(x)$, then the derivative at this point either is zero or does not exist. In other words, the extrema of a function are contained among its critical points.'

The proof of the necessary condition follows from Fermat's theorem.

The necessary condition does not guarantee the existence of an extremum. A classic illustration here is the cubic function $f(x) = x^3$. Despite the fact that the derivative of the function at the point $x=0$ is zero: $f'(x = 0) = 0$, this point is not an extremum.

Local extrema of differentiable functions exist when the sufficient conditions are satisfied. These conditions are based on the use of the first-, second-, or higher-order derivative. Respectively, 3 sufficient conditions for local extrema are considered. Now we turn to their formulation and proof.

First Derivative Test

Let the function $f(x)$ be differentiable in a neighborhood of the point x_0, except perhaps at the point x_0 itself, in which, however, the function is continuous.

Then,

- The derivative $f'(x)$ changes sign from minus to plus when passing through the point x_0 (from left to right), then x_0 is a strict minimum point. In other words, in this case there exists a number $\delta > 0$ such that,

$$\forall x \in (x_0 - \delta, x_0) \Rightarrow f'(x) < 0,$$
$$\forall x \in (x_0, x_0 + \delta) \Rightarrow f'(x) > 0,$$

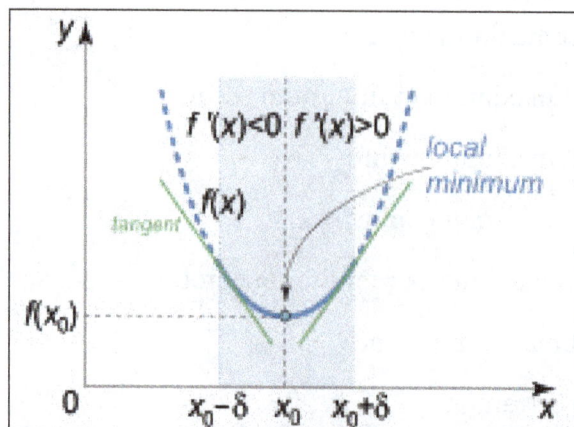

- If the derivative $f'(x)$, on the contrary, changes sign from plus to minus when passing through the point x_0, then x_0 is a strict maximum point. In other words, there exists a number $\delta > 0$ such that,

$$\forall x \in (x_0 - \delta, x_0) \Rightarrow f'(x) > 0,$$
$$\forall x \in (x_0, x_0 + \delta) \Rightarrow f'(x) < 0,$$

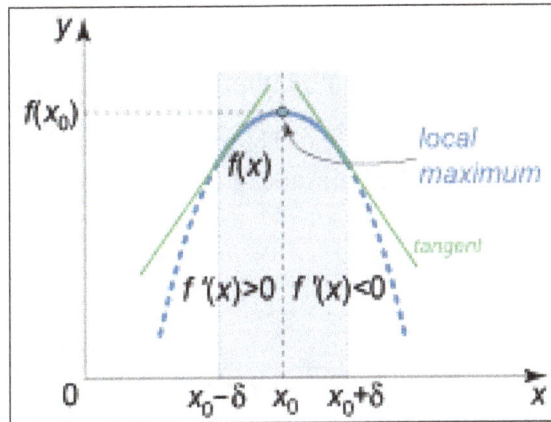

Proof: We confine ourselves to the case of the minimum. Suppose that the derivative $f'(x)$ changes sign from minus to plus when passing through the point x_0. To the left from the point x_0, the following condition is satisfied:

$$\forall x \in (x_0 - \delta, x_0) \Rightarrow f'(x) < 0.$$

By Lagrange's theorem, the difference of the values of the function at the points x and x_0 is written as,

$$f(x) - f(x_0) = f'(c)(x - x_0),$$

where the point c belongs to the interval $(x_0 - \delta, x_0)$, in which the derivative is negative, i.e. $f'(c) < 0$. Since $x - x_0 < 0$ to the left of the point x_0, then,

$$f(x) - f(x_0) > 0 \quad \text{for all } x \in (x_0 - \delta, x_0).$$

Likewise, it is established that:

$$f(x) - f(x_0) > 0 \quad \text{for all } x \in (x_0, x_0 + \delta).$$

to the right of the point x_0.

Based on the definition, we conclude that x_0 is a strict minimum point of the function $f(x)$. Similarly, we can prove the first derivative test for a strict maximum.

The first derivative test does not require the function to be differentiable at the point x_0. If the derivative at this point is infinite or does not exist (i.e. the point x_0 is critical, but not stationary), the first derivative test can still be used to investigate the local extrema of the function.

Second Derivative Test

Let the first derivative of a function $f(x)$ at the point x_0 be equal to zero: $f(x_0) = 0$, that is x_0 is a stationary point of $f(x)$. Suppose also that there exists the second derivative $f''(x_0)$ at this point.

Then,

- If $f''(x_0) > 0$, then x_0 is a strict minimum point of the function $f(x)$;

- If $f''(x_0) < 0$, then x_0 is a strict maximum point of the function $f(x)$.

Proof: In the case of a strict minimum $f''(x_0) > 0$. Then the first derivative is an increasing function at the point x_0. Consequently, there exists a number $\delta > 0$ such that,

$$\forall x \in (x_0 - \delta, x_0) \Rightarrow f'(x) > f'(x_0),$$
$$\forall x \in (x_0, x_0 + \delta) \Rightarrow f'(x) > f'(x_0),$$

Since $f''(x_0) = 0$ (because x_0 is a stationary point), therefore the first derivative is negative in the δ-neighborhood to the left of the point x_0, and is positive to the right, i.e. the derivative changes sign from minus to plus when passing through the point x_0. By the first derivative test, this means that x_0 is a strict minimum point.

The case of the maximum can be considered in a similar way.

The second derivative test is convenient to use when calculation of the first derivatives in the neighborhood of a stationary point is difficult. On the other hand, the second test may be used only for stationary points (where the first derivative is zero) – in contrast to the first derivative test, which is applicable to any critical points.

Third Derivative Test

Let the function $f(x)$ have derivatives at the point x_0 up to the nth order inclusively.

Then if,

$$f'(x_0) = f''(x_0) = \dots = f^{(n-1)}(x_0) = 0 \text{ and } f^{(n)}(x_0) \neq 0,$$

The point x_0 for even n is:

- A strict minimum point if $f^{(n)}(x_0) > 0$, and

- A strict maximum point if $f^{(n)}(x_0) > 0$.

For odd n, the extremum at x_0 does not exist.

It is clear that for $n = 2$, we obtain as a special case the second derivative test for local extrema considered above. To avoid such a transition, the third derivative test implies that $n > 2$.

Proof: Expand the function $f(x)$ at the point x_0 in a Taylor series,

$$f(x) = f(x_0) + \frac{f'(x_0)}{1!}(x-x_0) + \frac{f''(x_0)}{2!}(x-x_0)^2 + ...$$
$$+ \frac{f^{(n-1)}(x_0)}{(n-1)!}(x-x_0)^{n-1} + \frac{f^{(n)}(x_0)}{n!}(x-x_0)^n + o\left((x-x_0)^n\right)$$

Since, by assumption, all of the first derivatives up to the (n−1)th order are equal to zero, we obtain,

$$f(x) - f(x_0) = \frac{f^{(n)}(x_0)}{n!}(x-x_0)^n + o\left((x-x_0)^n\right),$$

where the remainder term o $((x - x_0)_n)$ has a higher order of smallness than n. As a result, the sign of the difference $f(x) - f(x_0)$ in the δ-neighborhood of the point x_0 will be determined by the sign of the nth term in the Taylor series,

$$sign\left[f(x) - f(x_0)\right] = sign\left[\frac{f^{(n)}(x_0)}{n!}(x-x_0)^n\right]$$

or

$$sign\left[f(x) - f(x_0)\right] = sign\left[f^{(n)}(x_0)(x-x_0)^n\right]$$

If n is an even number (n = 2k), then,

$$\forall x \in (x_0 - \delta, x_0 + \delta) \Rightarrow (x-x_0)^{2k} > 0$$

Consequently, in this case,

$$sign\left[f(x) - f(x_0)\right] = sign\ f^{(n)}(x_0).$$

If $f^{(n)}(x_0) > 0$ in the δ-neighborhood of the point x_0, then the following inequality holds:

$$f(x) - f(x_0) > 0.$$

By definition, this means that x_0 is a strict minimum point of the function $f(x)$. Similarly, if $f^{(n)}(x_0)$ < 0 in the δ-neighborhood of the point x_0, we have the inequality,

$$f(x) - f(x_0) < 0,$$

that corresponds to a strict maximum point.

If n is an odd number (n = 2k + 1), the degree of $(x - x_0)^{2k+1}$ will change sign when passing through the point x_0. Then it follows from the formula,

$$sign\left[f(x) - f(x_0)\right] = sign\left[f^{(n)}(x_0)(x-x_0)^{2k+1}\right]$$

that the difference $f(x) - f(x_0)$ also changes sign when passing through x_0. In this case, the extremum at the point x_0 does not exist.

Rules for Finding Derivatives

It is tedious to compute a limit every time we need to know the derivative of a function. Fortunately, we can develop a small collection of examples and rules that allow us to compute the derivative of almost any function we are likely to encounter. Many functions involve quantities raised to a constant power, such as polynomials and more complicated combinations like y = (sin x)⁴.

Power Rule

We start with the derivative of a power function, $f(x) = x^n$. Here n is a number of any kind integer rational, positive, negative, even irrational, as in x^π.

$$\frac{d}{dx}x^n = nx^{n-1}.$$

It is not easy to show this is true for any n. We will do some of the easier cases now, and discuss the rest later.

The easiest, and most common, is the case that n is a positive integer. To compute the derivative we need to compute the following limit:

$$\frac{d}{dx}x^n = \lim_{\Delta x \to 0} \frac{(x+\Delta x)^n - x^n}{\Delta x}.$$

For a specific, fairly small value of n, we could do this by straightforward algebra.

Example: Find the derivative of $f(x) = x^3$.

$$\frac{d}{dx}x^3 = \lim_{\Delta x \to 0} \frac{(x+\Delta x)^3 - x^3.}{\Delta x}$$

$$= \lim_{\Delta x \to 0} \frac{x^3 + 3x^2\Delta x + 3x\Delta x^2 + \Delta x^3 - x^3}{\Delta x}$$

$$= \lim_{x \to \infty} \frac{3x^2\Delta x + 3x\Delta x^2 + \Delta x^3}{\Delta x}$$

$$= \lim_{\Delta x \to \infty} 3x^2 + 3x\Delta x + \Delta x^2 = 3x^2.$$

The general case is really not much harder as long as we don't try to do too much. The key is understanding what happens when (x + Δx)n is multiplied out:

$$(x+\Delta x)^n = x^n + nx^{n-1}\Delta x + a_2 x^{n-2}\Delta x^2 + \text{.......} + + a_{n-1}x\Delta x^{n-1} + \Delta x^n.$$

We know that multiplying out will give a large number of terms all of the form $x^i \, \Delta x^j$, and in fact that $i + j = n$ in every term. One way to see this is to understand that one method for multiplying out $(x + \Delta x)^n$ is the following: In every $(x + \Delta x)$ factor, pick either the x or the x, then multiply the n choices together; do this in all possible ways. For example, for $(x + \Delta x)^3$, there are eight possible ways to do this,

$$
\begin{aligned}
(x + \Delta x)(x + \Delta x)(x + \Delta x) &= xxx + xx\Delta x + x\Delta xx + x\Delta x\Delta x \\
&\quad + \Delta xxx + \Delta xx\Delta x + \Delta x\Delta xx + \Delta x\Delta x\Delta x \\
&= x^3 + x^2\Delta x + x^2\Delta x + x\Delta x^2 \\
&\quad + x^2\Delta x + x\Delta x^2 + x\Delta x^2 + \Delta x^3 \\
&= x^3 + 3x^2\Delta x + 3x\Delta x^2 + \Delta x^3
\end{aligned}
$$

No matter what n is, there are n ways to pick Δx in one factor and x in the remaining $n-1$ factors; this means one term is $nx^{n-1}\Delta x$. The other coefficients are somewhat harder to understand, but we don't really need them, so in the formula above they have simply been called a_2, a_3, and so on. We know that every one of these terms contains Δx to at least the power 2. Now let's look at the limit:

$$
\begin{aligned}
\frac{d}{dx}x^n &= \lim_{\Delta x \to 0} \frac{(x + \Delta x)^n - x^n}{\Delta x} \\
&= \lim_{\Delta x \to 0} \frac{x^n + nx^{n-1}\Delta x + ax_2 x^{n-2}\Delta x^2 + \ldots + a_{n-1}x\Delta x^{n-1} + \Delta x^n - x^n}{\Delta x} \\
&= \lim_{\Delta x \to 0} \frac{nx^{n-1}\Delta x + a_2 x^{n-2}\Delta x^2 + \ldots + a_{n-1}x\Delta x^{n-1} + \Delta x^n}{\Delta x} \\
&= \lim_{\Delta x \to 0} nx^{n-1} + a_2 x^{n-2}\Delta x + \ldots + a_{n-1}x\Delta^{n-2} + \Delta x^{n-1}nx^{n-1}.
\end{aligned}
$$

Now without much trouble we can verify the formula for negative integers.

Example: Find the derivative of $y = x^{-3}$. Using the formula, $y' = -3x^{-3-1} = -3x^{-4}$.

Here is the general computation. Suppose n is a negative integer; the algebra is easier to follow if we use $n = -m$ in the computation, where m is a positive integer.

$$
\begin{aligned}
\frac{d}{dx}x^n = \frac{d}{dx}x^{-m} &= \lim_{\Delta x \to 0} \frac{(x + \Delta x)^{-m} - x^{-m}}{\Delta x} \\
&= \lim_{\Delta x \to 0} \frac{\dfrac{1}{(x + \Delta x)^m} - \dfrac{1}{x^m}}{\Delta x} \\
&= \lim_{\Delta x \to 0} \frac{x^m - (x + \Delta x)^m}{(x + \Delta x)^m \, x^m \Delta x} \\
&= \lim_{\Delta x \to 0} \frac{x^m - \left(x^m + mx^{m-1}\Delta x + a_2 x^{m-2}\Delta x^2 + \ldots + a_{m-1}x\Delta x^{m-1} + \Delta x^m\right)}{(x + \Delta x)^m \, x^m \Delta x}
\end{aligned}
$$

$$= \lim_{\Delta x \to 0} \frac{-mx^{m-1} - a_2 x^{m-2} \Delta x - \ldots a_{m-1} x \Delta x^{m-2} - \Delta x^{m-1}}{(x+\Delta x)^m x^m \Delta x}$$

$$= \frac{-mx^{m-1}}{x^m x^m} = \frac{-mx^{m-1}}{x^2 m} = -mx^{m-1} \Delta^{2m} = nx^{-m-1} = m = nx^{n-1}.$$

Let's note here a simple case in which the power rule applies, or almost applies, but is not really needed. Suppose that $f(x) = 1$; remember that this "1" is a function, not "merely" a number, and that $f(x) = 1$ has a graph that is a horizontal line, with slope zero everywhere. So we know that $f'(x) = 0$. We might also write $f(x) = x0$, though there is some question about just what this means at $x = 0$. If we apply the power rule, we get $f'(x) = 0x^{-1} = 0/x = 0$, again noting that there is a problem at $x = 0$. So the power rule "works" in this case, but it's really best to just remember that the derivative of any constant function is zero.

Linearity of the Derivative

An operation is linear if it behaves "nicely" with respect to multiplication by a constant and addition. The name comes from the equation of a line through the origin, $f(x) = mx$, and the following two properties of this equation. First, $f(cx) = m(cx) = c(mx) = cf(x)$, so the constant c can be "moved outside" or "moved through" the function f. Second, $f(x + y) = m(x + y) = mx + my = f(x) + f(y)$, so the addition symbol likewise can be moved through the function.

The corresponding properties for the derivative are,

$$\left(c f(x) \right)' = \frac{d}{dx} cf(x) = c \frac{d}{dx} f(x) cf'(x),$$

and

$$\left(f(x) + g(x) \right)' = \frac{d}{dx} \left(f(x) + g(x) \right) = \frac{d}{dx} f(x) + \frac{d}{dx} g(x) = f'(x) + g'(x).$$

It is easy to see, or at least to believe, that these are true by thinking of the distance/speed interpretation of derivatives. If one object is at position $f(t)$ at time t, we know its speed is given by $f'(t)$. Suppose another object is at position $5f(t)$ at time t, namely, that it is always 5 times as far along the route as the first object. Then it "must" be going 5 times as fast at all times.

The second rule is somewhat more complicated, but here is one way to picture it. Suppose a flatbed railroad car is at position $f(t)$ at time t, so the car is traveling at a speed of $f'(t)$ (to be specific, let's say that $f(t)$ gives the position on the track of the rear end of the car). Suppose that an ant is crawling from the back of the car to the front so that its position on the car is $g'(t)$ and its speed relative to the car is $g'(t)$. Then in reality, at time t, the ant is at position $f(t) + g(t)$ along the track, and its speed is "obviously" $f'(t) + g'(t)$.

Let's see how to verify these rules by computation. We'll do one and leave the other for the exercises.

$$\frac{d}{dx}\left(f(x)+g(x)\right)=\lim_{\Delta x\to 0}\frac{f(x+\Delta x)+g(x+\Delta x)-\left(f(x)+g(x)\right)}{\Delta x}$$

$$=\lim_{\Delta x\to 0}\frac{f(x+\Delta x)+g(x+\Delta x)-f(x)-g(x)}{\Delta x}$$

$$=\lim_{\Delta x\to 0}\frac{f(x+\Delta x)-f(x)+g(x+\Delta x)-g(x)}{\Delta x}$$

$$=\lim_{\Delta x\to 0}=\left(\frac{f(x+\Delta x)-f(x)}{\Delta x}+\frac{g(x+\Delta x)-g(x)}{\Delta x}\right)$$

$$=\lim_{\Delta x\to 0}\frac{f(x+\Delta x)-f(x)}{\Delta x}+\lim_{\Delta x\to 0}\frac{g(x+\Delta x)-g(x)}{\Delta x}$$

$$=f'(x)+g'(x)$$

This is sometimes called the sum rule for derivatives.

Example: Find the derivative of $f(x) = x^5 + 5x^2$. We have to invoke linearity twice here,

$$f'(x)=\frac{d}{dx}\left(x^5+5x^2\right)=\frac{d}{dx}x^5+\frac{d}{dx}\left(5x^2\right)=5x^4+5\frac{d}{dx}\left(x^2\right)=5x^4+5\cdot 2x^1=5x^4+10x.$$

Because it is so easy with a little practice, we can usually combine all uses of linearity into a single step. The following example shows an acceptably detailed computation.

Example: Find the derivative of $f(x)=3/x^4-2x^2+6x-7$.

$$f'(x)=\frac{d}{dx}\left(\frac{3}{x^4}-2x^2+6x-7\right)=\frac{d}{dx}\left(3x^{-4}-2x^2+6x-7\right)=-12x^{-5}-4x+6.$$

Product Rule

Consider the product of two simple functions, say $f(x) = (x^2 + 1)(x^3 - 3x)$. An obvious guess for the derivative of f is the product of the derivatives of the constituent functions: $(2x)(3x^2 - 3) = 6x^3 - 6x$. Is this correct? We can easily check, by rewriting f and doing the calculation in a way that is known to work. First, $f(x) = x^5 - 3x^3 + x^3 - 3x = x^5 - 2x^3 - 3x$, and then $f'(x) = 5x^4 - 6x^2 - 3$.

So the derivative of $f(x) g(x)$ is not as simple as $f'(x) g'(x)$. Surely there is some rule for such a situation? There is, and it is instructive to "discover" it by trying to do the general calculation even without knowing the answer in advance.

$$\frac{d}{dx}\left(f(x)g(x)\right)=\lim_{\Delta x\to 0}\frac{f(x+\Delta x)g(x+\Delta x)-f(x)g(x)}{\Delta x}$$

$$=\lim_{\Delta x\to 0}\frac{f(x+\Delta x)g(x+\Delta x)-f(x+\Delta x)g(x)+f(x+\Delta x)g(x)-f(x)g(x)}{\Delta x}$$

$$= \lim_{\Delta x \to 0} \frac{f(x+\Delta x)g(x+\Delta x)-f(x+\Delta x)g(x)}{\Delta x} + \lim_{\Delta x \to 0} \frac{f(x+\Delta x)g(x)-f(x)g(x)}{\Delta x}$$

$$= \lim_{\Delta x \to 0} f(x+\Delta x)\frac{g(x+\Delta x)-g(x)}{\Delta x} + \lim_{\Delta x \to 0} \frac{f(x+\Delta x)-f(x)}{\Delta x}g(x)$$

$$= f(x)g'(x)+f'(x)g(x)$$

A couple of items here need discussion. First, we used a standard trick, "add and subtract the same thing", to transform what we had into a more useful form. After some rewriting, we realize that we have two limits that produce $f'(x)$ and $g'(x)$. Of course, $f'(x)$ and $g'(x)$ must actually exist for this to make sense. We also replaced $\lim_{\Delta x \to 0} f(x+\Delta x)$ with $f(x)$—why is this justified?

What we really need to know here is that $\lim_{\Delta x \to 0} f(x+\Delta x) = f(x)$, r, that f is continuous at x. We already know that $f'(x)$ exists (or the whole approach, writing the derivative of f_g in terms of f' and g', doesn't make sense). This turns out to imply that f is continuous as well.

Here's why,

$$\lim_{\Delta x \to 0} f(x+\Delta x) = \lim_{\Delta x \to 0} \left(f(x+\Delta x)-f(x)+f(x) \right)$$

$$= \lim_{\Delta x \to 0} \frac{f(x+\Delta x)-f(x)}{\Delta x}\Delta x + \lim_{\Delta x \to 0} f(x)$$

$$= f'(x).0 + f(x) = f(x)$$

To summarize: the product rule says that:

$$\frac{d}{dx}(f(x)g(x)) = f(x)g'(x)+f'(x)g(x).$$

Returning to the example we started with, let $f(x) = (x^2 + 1)(x^3 - 3x)$. Then $f'(x) = (x^2 + 1)(3x^2 - 3) + (2x)(x^3 - 3x) = 3x^4 - 3x^2 + 3x^2 - 3 + 2x^4 - 6x^2 = 5x^4 - 6x^2 - 3$, as before. In this case it is probably simpler to multiply $f(x)$ out first, and then compute the derivative; here's an example for which we really need the product rule.

Example: Compute the derivative of $f(x) = x^2\sqrt{625-x^2}$.

We have already computed $\dfrac{d}{dx}\sqrt{625-x^2} = \dfrac{-x}{\sqrt{625-x^2}}$.

Now,

$$f'(x)x^2\frac{-x}{\sqrt{625-x^2}}+2x\sqrt{625-x^2} = \frac{-x^3+2x(625-x^2)}{\sqrt{625-x^2}} = \frac{-3x^3+1250x}{\sqrt{625-x^2}}.$$

Quotient Rule

What is the derivative of $(x^2 + 1)/(x^3 - 3x)$? More generally, we'd like to have a formula to compute the derivative of $f(x)/g(x)$ if we already know $f'(x)$ and $g'(x)$. Instead of attacking this problem head-on, let's notice that we've already done part of the problem: $f(x)/g(x) = f(x)\cdot(1/g(x))$, that is, this is "really" a product, and we can compute the derivative if we know $f'(x)$ and $(1/g(x))'$. So really the only new bit of information we need is $(1/g(x))'$ in terms of $g'(x)$.

As with the product rule, let's set this up and see how far we can get,

$$\frac{d}{dx}\frac{1}{g(x)} = \lim_{\Delta x \to 0} \frac{\frac{1}{g(x+\Delta x)} - \frac{1}{g(x)}}{\Delta x}$$

$$= \lim_{\Delta x \to 0} \frac{\frac{g(x)-g(x+\Delta x)}{g(x+\Delta x)g(x)}}{\Delta x}$$

$$= \lim_{\Delta x \to 0} \frac{g(x)-g(x+\Delta x)}{g(x+\Delta x)g(x)\Delta x}$$

$$= \lim_{x \to \infty} -\frac{g(x+\Delta x)-g(x)}{\Delta x}\frac{1}{g(x+\Delta x)g(x)}$$

$$= -\frac{g'(x)}{g(x)^2}$$

Now we can put this together with the product rule:

$$\frac{d}{dx}\frac{f(x)}{g(x)} = f(x)\frac{-g'(x)}{g(x)^2} + f'(x)\frac{1}{g(x)} = \frac{-f(x)g'(x)+f'(x)g(x)}{g(x)^2} = \frac{f'(x)g(x)-f(x)g'(x)}{g(x)^2}$$

Example: Compute the derivative of $(x^2+1)/(x^3-3x)$.

$$\frac{d}{dx}\frac{x^2+1}{x^3-3x} = \frac{2x(x^3-3x)-(x^2+1)(3x^2-3)}{(x^3-3x)^2} = \frac{-x^4-6x^2+3}{(x^3-3x)^2}.$$

It is often possible to calculate derivatives in more than one way, as we have already seen. Since every quotient can be written as a product, it is always possible to use the product rule to compute the derivative, though it is not always simpler.

Example: Find the derivative of $\sqrt{625-x^2}/\sqrt{x}$ in two ways: using the quotient rule, and using the product rule.

Quotient rule:

$$\frac{d}{dx}\frac{\sqrt{625-x^2}}{\sqrt{x}} = \frac{\sqrt{x}\left(-x/\sqrt{625-x^2}\right)-\sqrt{625-x^2}.1/\left(2\sqrt{x}\right)}{x}.$$

Note that we have used $\sqrt{x} = x^{1/2}$ to compute the derivative of $\sqrt{x}$ by the power rule.

Product rule:

$$\frac{d}{dx}\sqrt{625-x^2}\,x^{-1/2} = \sqrt{625-x^2}\,\frac{-1}{2}x^{-3/2} + \frac{-x}{\sqrt{625-x^2}}x^{-1/2} .$$

With a bit of algebra, both of these simplify to,

$$-\frac{x^2+625}{2\sqrt{625-x^2}\,x^{3/2}}$$

Occasionally you will need to compute the derivative of a quotient with a constant numerator, like $10/x^2$. Of course you can use the quotient rule, but it is usually not the easiest method. If we do use it here, we get,

$$\frac{d}{dx}\frac{10}{x^2} = \frac{x^2\cdot 0 - 10\cdot 2x}{x^4} = \frac{-20}{x^3},$$

since the derivative of 10 is 0. But it is simpler to do this:

$$\frac{d}{dx}\frac{10}{x^2} = \frac{d}{dx}10x^{-2} = -20x^{-3}.$$

Admittedly, x^2 is a particularly simple denominator, but we will see that a similar calculation is usually possible. Another approach is to remember that,

$$\frac{d}{dx}\frac{1}{g(x)} = \frac{-g'(x)}{g(x)^2},$$

but this requires extra memorization. Using this formula,

$$\frac{d}{dx}\frac{10}{x^2} = 10 - \frac{-2x}{x^4}$$

Note that we first use linearity of the derivative to pull the 10 out in front.

Chain Rule

So far we have seen how to compute the derivative of a function built up from other functions by addition, subtraction, multiplication and division. There is another very important way that we combine simple functions to make more complicated functions: function composition. For example, consider $\sqrt{625-x^2}$. This function has many simpler components, like 625 and x^2, and then there is that square root symbol, so the square root function $\sqrt{x} = x^{1/2}$ is involved. The obvious question is: can we compute the derivative using the derivatives of the constituents $625-x^2$ and $\sqrt{x}$? We can indeed. In general, if $f(x)$ and $g(x)$ are functions, we can compute the derivatives of $f(g(x))$ and $g(f(x))$ in terms of $f'(x)$ and $g'(x)$.

Example: Form the two possible compositions of $f(x) = \sqrt{x}$ and $g(x) = 625 - x^2$ and compute the derivatives. First, $f(g(x)) = \sqrt{625 - x^2}$, and the derivative is $-x\sqrt{625 - x^2}$ as we have seen. Second, $g(f(x)) = 625 - \left(\sqrt{x}\right)^2 = 625 - x$ with derivative -1. Of course, these calculations do not use anything new, and in particular the derivative of $f(g(x))$ was somewhat tedious to compute from the definition.

Suppose we want the derivative of $f(g(x))$. Again, let's set up the derivative and play some algebraic tricks,

$$\frac{d}{dx} f\left(g(x)\right) = \lim_{\Delta x \to 0} \frac{f\left(g(x+\Delta x)\right) - f\left(g(x)\right)}{\Delta x}$$
$$= \lim_{\Delta x \to 0} \frac{f\left(g(x+\Delta x)\right) - f\left(g(x)\right)}{g(x+\Delta x) - g(x)} \frac{g(x+\Delta x) - g(x)}{\Delta x}$$

Now we see immediately that the second fraction turns into $g'(x)$ when we take the limit. The first fraction is more complicated, but it too looks something like a derivative. The denominator, $g(x + \Delta x) - g(x)$, is a change in the value of g, so let's abbreviate it as,

$$\Delta g = g\left(x+\Delta x\right) - g\left(x\right),$$

Which also means $g\left(x+\Delta x\right) = g\left(x+\Delta x\right) = g\left(x\right) + \Delta g$. This gives us:

$$\lim_{\Delta x \to 0} \frac{f\left(g(x)+\Delta g\right) - f\left(g(x)\right)}{\Delta g}.$$

As Δx goes to 0, it is also true that Δg goes to 0, because $g(x + \Delta x)$ goes to $g(x)$. So we

can rewrite this limit as:

$$\lim_{\Delta g \to 0} \frac{f\left(g(x)+\Delta g\right) - f\left(g(x)\right)}{\Delta g}.$$

Now this looks exactly like a derivative, namely $f'(g(x))$, that is, the function $f'(x)$ with x replaced by $g(x)$. If this all withstands scrutiny, we then get:

$$\frac{d}{dx} f\left(g(x)\right) = f'\left(g(x)\right) g'(x).$$

Unfortunately, there is a small flaw in the argument. Recall that what we mean by $\lim_{\Delta x \to 0}$ involves what happens when Δx is close to 0 but not equal to 0. The qualification is very important, since we must be able to divide by Δx. But when Δx is close to 0 but not equal to 0, $\Delta g = g(x + \Delta x))$ $\Delta g(x)$ is close to 0 and possibly equal to 0. This means it doesn't really make sense to divide by Δg. Note that many functions g do have the property that $g(x + \Delta x) - g(x) \neq = 0$ when Δx is small, and for these functions the argument above is fine.

The chain rule has a particularly simple expression if we use the Leibniz notation for the derivative.

The quantity $f'(g(x))$ is the derivative of f with x replaced by g; this can be written df/dg. As usual, $g'(x)$ = dg/dx. Then the chain rule becomes:

$$\frac{df}{dx} = \frac{df}{dg}\frac{dg}{dx}.$$

This looks like trivial arithmetic, but it is not: dg/dx is not a fraction, that is, not literal division, but a single symbol that means $g'(x)$. Nevertheless, it turns out that what looks like trivial arithmetic, and is therefore easy to remember, is really true.

It will take a bit of practice to make the use of the chain rule come naturally—it is more complicated than the earlier differentiation rules we have seen.

Example: Compute the derivative of $\sqrt{625-x^2}$. We already know that the answer is $-x\sqrt{625-x^2}$, computed directly from the limit. In the context of the chain rule, we have $f(x)=\sqrt{x}, g(x)=625-x^2$. We know that $f'(x)$ = $(1/2)x^{-1/2}$, so $f'(g(x))=(1/2)(625-x^2)^{-1/2}$. Note that this is a two-step computation: first compute $f'(x)$, then replace x by $g(x)$.

Since $g'(x)$ = $-2x$ we have:

$$f'(g(x))g'(x) = \frac{1}{2\sqrt{625-x^2}}(-2x) = \frac{-x}{\sqrt{625-x^2}}.$$

Example: Compute the derivative of $1/\sqrt{625-x^2}$. This is a quotient with a constant numerator, so we could use the quotient rule, but it is simpler to use the chain rule. The function is $\left(625-x^2\right)^{-1/2}$ the composition of $f(x)$ = $x^{-1/2}$ and $g(x)$ = 625 − x2. We compute $f'(x)$ = $(-1/2)x^{-3/2}$ using the power rule, and then:

$$f'(g(x))g'(x) = \frac{-1}{2\left(625-x^2\right)^{3/2}}(-2x) = \frac{x}{\left(625-x^2\right)^{3/2}}.$$

In practice, of course, you will need to use more than one of the rules we have developed to compute the derivative of a complicated function.

Example: Compute the derivative of:

$$f(x) = \frac{x^2-1}{x\sqrt{x^2+1}}.$$

The "last" operation here is division, so to get started we need to use the quotient rule first. This gives,

$$f'(x) = \frac{\left(x^2-1\right)'x\sqrt{x^2+1}-\left(x^2-1\right)\left(x\sqrt{x^2+1}\right)'}{x^2\left(x^2+1\right)}$$

$$= \frac{2x^2\sqrt{x^2+1} - (x^2-1)(x\sqrt{x^2+1})'}{x^2(x^2+1)}$$

Now we need to compute the derivative of $x\sqrt{x^2+1}$. This is a product, so we use the product rule:

$$\frac{d}{dx}x\sqrt{x^2+1} = x\frac{d}{dx}\sqrt{x^2+1} + \sqrt{x^2+1}.$$

Finally, we use the chain rule:

$$\frac{d}{dx}\sqrt{x^2+1} = \frac{d}{dx}(x^2+1)^{1/2} = \frac{1}{2}(x^2+1)^{-1/2}(2x) = \frac{x}{\sqrt{x^2+1}}.$$

And putting it all together:

$$f'(x) = \frac{2x^2\sqrt{x^2+1} - (x^2-1)(x\sqrt{x^2+1})'}{x^2(x^2+1)}$$

$$= \frac{2x^2\sqrt{x^2+1} - (x^2-1)\left(x\dfrac{x}{\sqrt{x^2+1}} + \sqrt{x^2+1}\right)}{x^2(x^2+1)}.$$

This can be simplified of course, but we have done all the calculus, so that only algebra is left.

Example: Compute the derivative of $\sqrt{1+\sqrt{1+\sqrt{x}}}$. Here we have a more complicated chain of compositions, so we use the chain rule twice. At the outermost "layer" we have the function $g(x) = 1 + \sqrt{1+\sqrt{x}}$ plugged into $f(x) = \sqrt{x}$.

So applying the chain rule once gives:

$$\frac{d}{dx}\sqrt{1+\sqrt{1+\sqrt{x}}} = \frac{1}{2}\left(1+\sqrt{1+\sqrt{x}}\right)^{-1/2}\frac{d}{dx}\left(1+\sqrt{1+\sqrt{x}}\right).$$

Now we need the derivative of $\sqrt{1+\sqrt{x}}$. Using the chain rule again,

$$\frac{d}{dx}\sqrt{1+\sqrt{x}} = \frac{1}{2}\left(1+\sqrt{x}\right)^{-1/2}\frac{1}{2}x^{-1/2}$$

So the original derivative is,

$$\frac{d}{dx}\sqrt{1+\sqrt{1+\sqrt{x}}} = \frac{1}{2}\left(1+\sqrt{1+\sqrt{x}}\right)^{-1/2}\frac{1}{2}\left(1+\sqrt{x}\right)^{-1/2}\frac{1}{2}x^{-1/2}.$$

$$= \frac{1}{8\sqrt{x}\sqrt{1+\sqrt{x}}\sqrt{1+\sqrt{1\sqrt{x}}}}$$

Using the chain rule, the power rule, and the product rule, it is possible to avoid using the quotient rule entirely.

Example: Compute the derivative of $f(x) = \dfrac{x^3}{x^2+1}$. Write $f(x) = x^3 (x^2 + 1)^{-1}$,

$$f'(x) = x^3 \frac{d}{dx}\left(x^2+1\right)^{-1} + 3x^2\left(x^2+1\right)^{-1}$$

$$= x^3(-1)\left(x^2+1\right)^{-2}(2x) + 3x^2\left(x^2+1\right)^{-1}$$

$$= -2x^4\left(x^2+1\right)^{-2} + 3x^2\left(x^2+1\right)^{-1}$$

$$= \frac{-2x^4}{\left(x^2+1\right)^2} + \frac{3x^2}{x^2+1}$$

$$= \frac{-2x^4}{\left(x^2+1\right)^2} + \frac{3x^2\left(x^2+1\right)}{\left(x^2+1\right)^2}$$

$$= \frac{-2x^4 + 3x^4 + 3x^2}{\left(x^2+1\right)^2} = \frac{x^4 + 3x^2}{\left(x^2+1\right)^2}$$

We already had the derivative on the second line; all the rest is simplification. It is easier to get to this answer by using the quotient rule.

Rates of Change and Applications to Motion

Average Rates of Change

Suppose $s(t) = 2t^3$ represents the position of a race car along a straight track, measured in feet from the starting line at time t seconds. What is the average rate of change of $s(t)$ from $t = 2$ to $t = 3$?

The average rate of change is equal to the total change in position divided by the total change in time,

$$\text{Avg Rate} = \frac{\Delta s}{\Delta t}$$

$$= \frac{s(3) - s(2)}{3 - 2}$$

$$= \frac{54 - 16}{1}$$

$$= 38\,\text{ft per second}$$

In physics, velocity is the rate of change of position. Thus, 38 feet per second is the average velocity of the car between times t = 2 and t = 3.

Instantaneous Rates of Change

What is the instantaneous rate of change of the same race car at time t = 2? The instantaneous rate of change measures the rate of change, or slope, of a curve at a certain instant. Thus, the instantaneous rate of change is given by the derivative. In this case, the instantaneous rate is s'(2).

$$s'(t) = 6t^2$$

$$s'(2) = 6(2)^2 = 24 \text{ feet per second}$$

Thus, the derivative shows that the race car had an instantaneous velocity of 24 feet per second at time t = 2.

Rectilinear Motion

Rectilinear Motion refers to the motion of an object in a straight line. Such motion can be depicted as a point which moves forwards and/or backwards on a number line.

General Motion Equations

If s(t) represents the position of the object on the number line at time t, then v(t), the (instantaneous) velocity, is equal to s'(t), and a(t), the (instantaneous) acceleration, is equal to v'(t), which is s"(t).

Thus, velocity is the rate of change of position, and acceleration is the rate of change of velocity.

Example: If $s(t) = t^2 - 5t$, what is the position, velocity and acceleration at $t = 2$? Assume s is in feet and t is in seconds, and interpret these results.

$$s(t) = t^2 - 5t + 3$$

$$v(t) = s'(t) = 2t - 5$$

$$a(t) = v'(t) = 2$$

$$s(2) = 2$$

$$v(2) = -1$$

$$a(2) = 2.$$

So, at t = 2, the object is located at +2 feet from the start. The velocity is -1 foot per second. The negative sign indicates that it is headed in the negative direction, and it is moving backwards at a rate of one foot per second. The acceleration is 2, which means that at that instant, its velocity is increasing by a rate of 2 feet per second each second.

Vector and Scalar Quantities

Position, velocity, and acceleration are all vector quantities because they contain both a direction and a magnitude. For example, if the velocity of an object is -3 feet per second, then that object is moving backwards (direction) at a rate of 3 feet per second (magnitude). Similarly, if an object has a position of -3 feet, then is located 3 feet from the starting point (magnitude), but on the negative side (direction).

The vector quantities of position and velocity both have corresponding scalar quantities that only have a magnitude. The scalar analog of position is distance. Although the position of an object with respect to the start line may be -3 feet, its distance from that start line is simply 3 feet, because distance is always a positive quantity. Thus, distance is the absolute value of position.

Similarly, the scalar analog of velocity is speed. Whether an object's velocity is -5 feet per second, or + 5 feet per second, its speed is still simply 5 feet per second, because speed is always a positive quantity that contains no information about direction. Thus, speed is the absolute value of velocity.

Motion with Constant Acceleration

This is the special case of rectilinear motion in which the acceleration is constant. In cases where the acceleration is constant, a(t) can be represented simply by the constant a, and both velocity and position can be found by using the following formulas,

$$a(t) = a$$
$$v(t) = v_0 + at$$
$$s(t) = s_0 + v_0 t + \frac{1}{2} + at^2$$

where v_0 is the initial velocity at time $t = 0$ and s_0 is the initial position at time $t = 0$. Note that these formulas are in compliance with the relations $v(t) = s'(t)$ and $a(t) = v'(t)$.

A ball dropped vertically from a height travels in this fashion, because it is accelerated by gravity at a constant rate of 9.8 meters per second per second.

References

- Derivative-mathematics, science: britannica.com, Retrieved 2 February, 2019

- The-derivative: hec.ca, Retrieved 4 March, 2019

- Local-extrema-functions: math24.net, Retrieved 20 January, 2019

- Calculus, Rules-for-Finding-Derivatives, mathematics: whitman.edu, Retrieved 27 August, 2019

- Applicationsofthederivative, math: sparknotes.com, Retrieved 8 April, 2019

The Integral

Integral is an operation of calculus which assigns numbers to functions such that it describes displacement, area, volume, and other concepts which occur due to combination of infinitesimal data. The various techniques used within integration are integration by substitution, integration by parts and integration by trogonometric substitution. This chapter discusses in detail these techniques of integration.

In Calculus, Integral is either a numerical value equal to the area under the graph of a function for some interval (definite integral) or a new function the derivative of which is the original function (indefinite integral). These two meanings are related by the fact that a definite integral of any function that can be integrated can be found using the indefinite integral and a corollary to the fundamental theorem of calculus. The definite integral (also called Riemann integral) of a function $f(x)$ is denoted as,

$$\int_a^b f(x)\,dx$$

and is equal to the area of the region bounded by the curve (if the function is positive between $x = a$ and $x = b$) $y = f(x)$, the x-axis, and the lines $x = a$ and $x = b$. An indefinite integral, sometimes called an antiderivative, of a function $f(x)$, denoted by,

$$\int f(x)\,dx.$$

is a function the derivative of which is $f(x)$. Because the derivative of a constant is zero, the indefinite integral is not unique. The process of finding an indefinite integral is called integration.

The Indefinite Integral

The indefinite integral (also called the antiderivative, and sometimes the primitive integral) is related to the definite integral through the fundamental theorem of calculus. We know that the definite integral will give us the area of the region under a curve for a continuous function over a closed interval. As such, it evaluates to a number. The indefinite integral does not evaluate to a number. Rather, the indefinite integral is a function. In fact, it is an entire family of functions with an infinite number of members.

The name antiderivative actually describes the nature of the indefinite integral quite well. It is, in essence, the opposite of the derivative. Suppose we have a function $f(x)$ for which we want to find the indefinite integral. We have already established that we are looking for a function. It turns out that $f(x)$ is actually the derivative of the function we are looking for. Let's look at an example. Suppose we want to find the indefinite integral of the function $f(x) = x^2 + 2x$. We know that $f(x)$ is the derivative of the function we are looking for, but how do we reverse the process of differentiation in order to get the indefinite integral? Let's think about what we had to do to get the derivative in the first place.

Consider the term x^2. When we differentiate a power of x, we multiply the coefficient of x by the exponent to which x is raised, and then decrement the exponent by one. To reverse the process, we need to increment the exponent by one, and divide the coefficient of x by the new exponent. The new exponent of x will therefore be three, and the new coefficient of x will be one third ($^1/_3$). The integral of x^2 is thus $^1/_3 x^3$. Applying the same procedure to the term 2x, we see that the integral of 2x must be x^2.

The function we are looking for - which we'll call F(x) - will therefore be,

$$F(x) = 1/3\,x^3 + x^2$$

If this is correct, then finding the derivative of this function should give us the original function $f(x) = x^2 + 2x$. Applying the basic rules of differentiation to F(x) will confirm that this is the case,

$$F'(x) = x^2 + 2x$$

In general terms, we can define the indefinite integral of $f(x)$ as any function F(x) such that,

$$F'(x) = f(x)$$

There is a potential problem here, however. Consider the following possibilities for F(x),

- $F(x) = ^1/_3 x^3 + x^2 + 5$

- $F(x) = ^1/_3 x^3 + x^2 + \sqrt{86}$

- $F(x) = ^1/_3 x^3 + x^2 + \pi$

- $F(x) = ^1/_3 x^3 + x^2 + e^2$

Now think about what the derivative will be for each of these functions. You will of course find that they will all have the same derivative, i.e.

$$F'(x) = x^2 + 2x$$

Because the last term in every case is a constant. Whenever we differentiate a constant, we get zero. It therefore doesn't really matter what constant term we have at the end of a function when differentiating. Any collection of functions that only differ from one another by a constant term will all have the same derivative. When you think about it, this is perfectly logical. Because the derivative of a function simply gives us the slope of that function for a given value of x. Adding a constant value to a function does not change its slope, merely its vertical orientation.

There are in fact an infinite number of functions that will give us exactly the same derivative. The only difference between them will be the constant term.

We could therefore perhaps write the indefinite integral of the function $f(x) = x^2 + 2x$ as follows:

$$F(x) = ^1/_3 x^3 + x^2 + ?$$

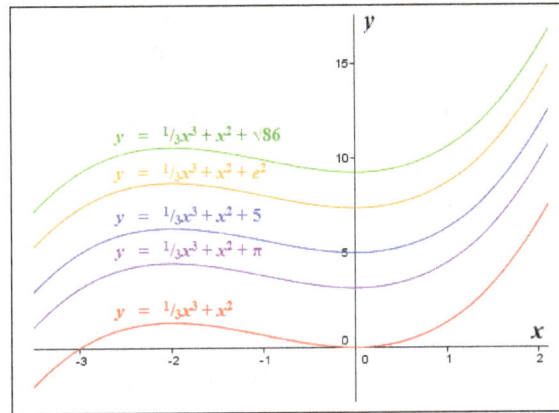

Adding a constant term to a function does not alter its slope.

Where the question mark represents the unknown constant value. We'll come back to the question of how we deal with this unknown value shortly. Meanwhile, let's turn our attention to the notation we should be using here for the indefinite integral. You probably remember how to write a definite integral for a function $f(x)$:

$$\int_a^b f(x)\,dx.$$

The integral symbol (that's the long 's' character over on the left-hand side) tells us that we are looking at an integral, and the subscripted and superscripted characters a and b immediately to the right of the integration symbol are the lower and upper limits of integration respectively. The function $f(x)$ is the integrand (i.e. the thing we are integrating), and the dx at the end tells us that x is our variable of integration (it can also be seen as representing infinitesimally small increments of x). Now look at how we write the indefinite integral of $f(x)$:

$$\int f(x)\,dx.$$

At first glance, this looks the same as the notation for a definite integral. Note, however, that the upper and lower limits of integration are missing. This is because there is no domain of integration. Whereas the definite integral leads us to a number that represents the area of a bounded region under the graph of a function, the indefinite integral is simply another function - the function we get, in fact, by reversing the process of differentiation that gave us the function $f(x)$. This process, which is the inverse of differentiation, is called antidifferentiation (or indefinite integration).

We still need to do something about the constant term that was lost when we carried out our differentiation (to simplify things, we will work on the assumption that there was a constant term, even if there wasn't). There is of course no way to determine the value of the constant term. Once it has been eliminated in the differentiation process, it is gone for good. But how do we show this missing constant in our notation? The answer is actually very simple. We just use the letter C as a placeholder. Here is how we write the indefinite integral of the function $f(x) = x^2 + 2x$:

$$\int x^2 + 2x\,dx = 1/3\ x^3 + x^2 + C.$$

The letter C represents all possible values of the missing constant, including zero. We call C the constant of integration. The dx that follows the integrand is a differential. In fact, you should already be quite familiar with it from your study of differential calculus. It is very important, and must never be omitted. Why? Well for one thing, it tells us where the expression to be integrated (the integrand) ends. In the above example, its absence would not really cause a problem, but consider the following indefinite integral:

$$\int 3x^2 + 2x \; dx + 3 = x^3 + x^2 + C + 3.$$

Suppose we forgot to include the dx on the left-hand side of the equation? Obviously, our notation would be incorrect. Of far more importance, though, is the fact that we would get the wrong answer, i.e.

$$\int 3x^2 + 2x + 3 = x^3 + x^2 + 3x + C.$$

There are other good reasons for always including the differential. For example, it tells us which variable we are integrating for (i.e. whether we are integrating for x, or for some other variable). This is particularly important if we wish to venture into the realms of multivariable calculus. Even if we never go further than single-variable calculus, we will inevitably encounter integration problems far more complex than the examples we have looked at so far. This will often involve manipulating equations using algebraic operations that rely on the presence of the differential.

Putting everything we have learnt so far together, we can now express the indefinite integral of a function $f(x)$ as follows,

$$\int f(x)\,dx = F(x) + C$$

where F(x) satisfies the condition that:

$$F'(x) = f(x).$$

Although we will talk about the fundamental theorem of calculus in much more detail elsewhere, it is worth briefly outlining the theorem here in order to give you an idea of how integration and differentiation are related, and why the indefinite integral can help us to calculate a definite integral for an integrable function. The theorem itself is in two parts. The first part, which is sometimes called the first fundamental theorem of calculus, essentially just tells us that integration and differentiation are the inverse of one another. The second part, which is sometimes called the second fundamental theorem of calculus, tells us that we can calculate a definite integral for a function using one of its indefinite integrals (of which, remember, there are an infinite number).

We have already seen an example of how we can apply the first part of the theorem to find the indefinite integral of a function, simply by applying the power rule for integration. This rule is essentially the inverse of the power rule used in differentiation. It is based on Cavalieri's quadrature formula, which is named after the seventeenth century Italian mathematician Bonaventura Francesco Cavalieri. The integration power rule formula is as follows:

$$\int ax^n dx = a\frac{x^{n+1}}{n+1} + C.$$

Note that n must not be equal to minus one (n ≠ −1) because this would put a zero in the denominator on the right hand side of the formula. The letters a and C represent the constant coefficient of x, and the constant of integration respectively. This rule on its own enables us to integrate all polynomial functions of one variable. As with differentiation, we integrate each term separately, and the plus or minus sign in front of each term does not change. More complex functions will require additional rules. Integration has no equivalents for the product and quotient rules used in differentiation. If we encounter a product or a quotient in an integration problem, we need to find other ways of dealing with them.

The most important consequence of the second fundamental theorem of calculus is that it gives us a relatively straightforward way of evaluating a definite integral for a function. In a nutshell, it tells us that if a function is continuous over some closed interval, then the definite integral for that interval (or domain of integration) can be calculated by finding the values of the indefinite integral (which is a function, remember) at each end of the interval. The definite integral will be the difference between these two values. In other words, if the function F(x) is the indefinite integral of the function f(x), and f(x) is continuous over the closed interval [a, b], then:

$$\int_a^b f(x)\,dx = F(b) - F(a).$$

Let's try an example. We'll find the definite integral for the function f(x) = 2x⁵ - 10x³ + 5 for -2 ≤ x ≤ 2. The graph of this function is shown below. The definite integral will have both positive and negative components.

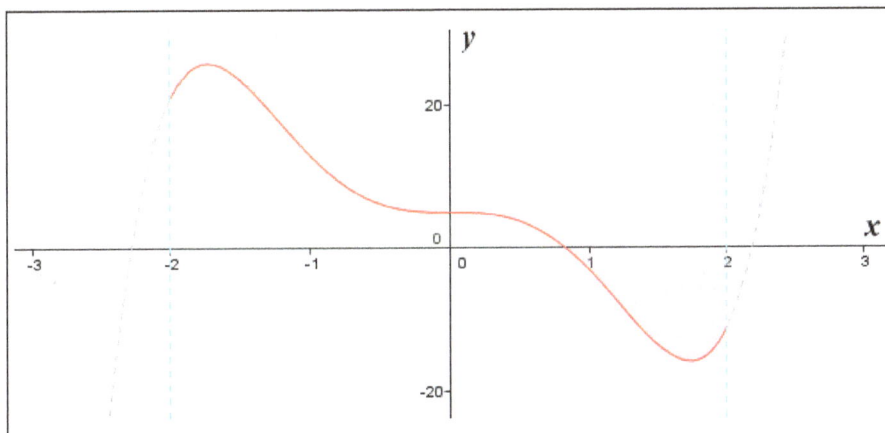

The graph of the function f(x) = 2x⁵ - 10x³ + 5 for -2 ≤ x ≤ 2.

Applying the power rule for integration to each term of the function in turn, we get the following:

$$\int 2x^5\,dx = 2\frac{x^6}{6} = \frac{1}{3}x^6$$

$$\int -10x^3\,dx = -10\frac{x^4}{4} = -\frac{5}{2}x^4$$

$$\int 5\,dx = 5\frac{x}{1} = 5x$$

The indefinite integral F(x) for the function $f(x) = 2x^5 - 10x^3 + 5$ is therefore given by:

$$F(x) = \int 2x^5 - 10x^3 + 5 \, dx = 1/3 x^6 - 5/2 x^4 + 5x.$$

And the definite integral for the function $f(x) = 2x^5 - 10x^3 + 5$ for $-2 \le x \le 2$ will be:

$$\int_{-2}^{2} 2x^5 - 10x^3 + 5 \, dx = F(2) - f(-2)$$

$$= (21^1/_3 - 40 + 10) - (21^1/_3 - 40 - 10)$$

$$= -8^2/_3 + 28^2/_3 = 20$$

Basic Properties of Indefinite Integrals

Since integration is the reverse process of differentiation, we can derive some basic properties of indefinite integrals from differentiation.

- $\int kf(x) \, dx = k \int f(x) \, dx$, Where k is a constant.

- $\int f(x) \pm g(x) \, dx = \int f(x) \, dx \pm \int g(x) \, dx.$

Proof: Suppose F(x) is the antiderivative of f(x), i.e., $\dfrac{d}{dx} F(x) = f(x)$.

$$\frac{d}{dx}\left[kF(x) \right] = k \frac{d}{dx} F(x)$$

$$= kf(x)$$

By the definition of indefinite integral:

$$\int kf(x) \, dx = kF(x) + C_1.$$

On the other hand:

$$k \int f(x) \, dx = k \left[F(x) + C \right]$$

$$= kF(x) + C_2.$$

Since C_1 and C_2 are arbitary constants:

$$\int kf(x) \, dx = k \int f(x) \, dx.$$

Proof: Suppose F(x) and G(x) are the antiderivative of f(x) and g(x) respectively.

i.e., $\dfrac{d}{dx} G(x) = g(x).$

$$\frac{d}{dx}\left[F(x) \pm G(x) \right] = \frac{d}{dx} F(x) \pm \frac{d}{dx} G(x)$$

$$= f(x) \pm g(x)$$

By the definition of indefinite integral:

$$\int \left[f(x) \pm g(x) \right] dx = F(x) \pm G(x) + C$$
$$= F(x) + C_1 \pm G(x) + C_2$$
$$= \int f(x) dx \pm \int g(x) dx.$$

Basic Integrals-elementary Integrals

1. $\int 1 dx = x + C$

2. $\int x \, dx = \frac{1}{2} x^2 + C$

3. $\int x^2 dx = \frac{1}{3} x^3 + C$

4. $\int \frac{1}{x^2} dx = -\frac{1}{x} + C$

5. $\int \sqrt{x} \, dx = \frac{2}{3} x^{3/2} + C$

6. $\int \frac{1}{\sqrt{x}} dx = 2\sqrt{x} + C$

7. $\int x^r dx = \frac{1}{r+1} x^{r+1} + C \ (r \neq -1)$

8. $\int \frac{1}{x} dx = \ln |x| + C$

9. $\int \sin \, ax \, dx = -\frac{1}{a} \cos ax + C$

10. $\int \cos \, ax \, dx = \frac{1}{a} \sin \, ax + C$

11. $\int \sec^2 ax \, dx = \frac{1}{a} \tan \, ax + C$

12. $\int esc^2 ax \, dx = -\frac{1}{a} cse \, ax + C$

13. $\int \sec^2 \, ax \, dx = \frac{1}{a} \tan ax + C$

14. $\int cse \, ax \, \cot \, ax \, dx = -\frac{1}{a} cse \, ax + C$

15. $\int \frac{1}{\sqrt{a^2 - x^2}} dx = \sin^{-1} \frac{x}{a} + C \, (a > 0)$

16. $\int \frac{1}{a^2 + x^2} = \frac{1}{a} \tan^{-1} \frac{x}{a} + C$

17. $\int e^{ax} \, dx = \frac{1}{a} e^{ax} + C$

18. $\int b^{ax} dx \ \frac{1}{a \, in \, b} b^{ax} + C$

19. $\int Cosh \, ax \, dx = \frac{1}{a} \sin h \, ax + C$

20. $\int \sinh \, ax \, dx = \frac{2}{a} \cosh \, ax + C$

Example: Find $\int \left(2x - \frac{3}{\sqrt{x}} \right) dx.$

$$\int \left(2x - \frac{3}{\sqrt{x}} \right) dx = 2 \int x \, dx - 3 \int x^{-\frac{1}{2}} dx$$

$$= 2 \left(\frac{x^2}{2} \right) - 3 \left(\frac{x^{\frac{1}{2}}}{\frac{1}{2}} \right) + C$$

$$= x^2 - 6x^{\frac{1}{2}} + C$$

Example: Find $\int \frac{x^2+6}{x^2}dx$.

$$\int \frac{x^2+6}{x^2}dx = \int\left(1+\frac{6}{x^2}\right)dx$$
$$= \int\left(1+6x^{-2}\right)dx$$
$$= x-\frac{6}{x}+C$$

Example: Find $\int \sec x(\cos x+\tan x)dx$.

$$\int \sec x(\cos x+\tan x)dx = \int(\sec x\cos x\tan x)dx$$
$$= \int(1+\sec x\tan x)dx$$
$$= x+\sec x+C$$

Example: Find $\int\left(6x^2+\frac{1}{x}-e^x\right)dx$.

$$\int\left(6x^2+\frac{1}{x}-e^x\right)dx = 6\left(\frac{x^3}{3}\right)+In|x|-e^x+C$$
$$= 2x^3+In|x|e^x+C$$

Example: Find $\int \frac{1}{1-\sin^2 x}dx$.

$$\int \frac{1}{1-\sin^2 x}dx = \int \frac{1}{\cos^2 x}dx$$
$$= \int \sec^2 x\, dx$$
$$= \tan x+C$$

Example: Find $\int \frac{1+\cos\theta}{1-\cos\theta}d\theta$

$$\int \frac{1+\cos\theta}{1-\cos\theta}d\theta = \int \frac{1+\cos\theta}{1+\cos\theta}\times\frac{1+\cos\theta}{1+\cos\theta}d\theta$$
$$= \int \frac{1+2\cos\theta+\cos^2\theta}{\sin^2\theta}d\theta$$
$$= \int\left(\csc^2\theta+2\cot\theta\csc\theta+\cot^2\theta\right)d\theta$$
$$= -\cot\theta-2\csc\theta+\int\left(\csc^2\theta-1\right)d\theta$$
$$= -2\cot\theta-2\csc\theta-\theta+C$$

The Definite Integral

If a function is continuous over some defined interval, it is said to be integrable, because we can find its definite integral. Suppose we have a continuous function $f(x)$ defined on an interval [a, b]. We can informally define the definite integral as the signed area of the region bounded by the graph of $f(x)$, the x axis, and the vertical lines intersecting the x axis at a and b. The area of any part of the region above the x axis is positive, while the area of any part of the region below the x axis is negative. The definite integral is the net area between a function and the x axis.

There are various methods we could use to get an approximation of the area. The illustration below shows the function $f(x) = x^2$ defined on the interval [0, 1]. The region under the curve is divided into eight subintervals of equal width. The rectangles that we see inscribed within each subinterval have the same width as the subinterval, and a height equal to the smallest value taken by $f(x)$ over the subinterval.

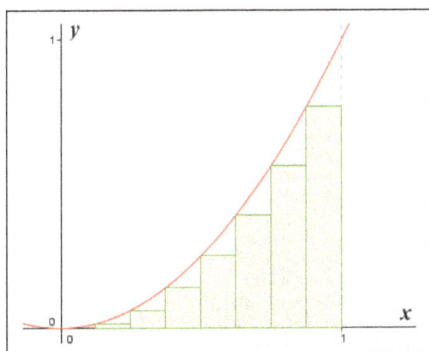

Approximating the area under $f(x) = x^2$ for $0 \le x \le 1$ using inscribed rectangles.

If we add together the areas of the rectangles, we will get the left end point approximation for the area of the region under the curve (so called because each rectangle is drawn so that its top left-hand corner lies on the curve of the graph). Because the function is increasing over the interval, none of the rectangles cover the entire region under the graph contained within their subinterval. The approximation will therefore be an underestimate of the actual area. Suppose, for the sake of comparison, we make another approximation using rectangles. Everything will be exactly the same, except that this time we'll make the height of the rectangle drawn within each subinterval equal to the largest value taken by $f(x)$ over the subinterval.

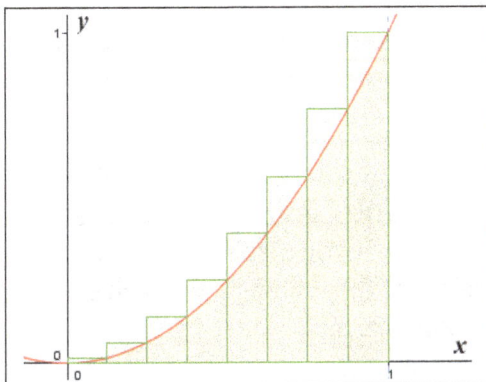

Approximating the area under $f(x) = x^2$ for $0 \le x \le 1$ using circumscribed rectangles.

This time, when we add together the areas of the rectangles, we get the right end point approximation for the area of the region under the curve (so called because each rectangle is drawn so that its top right-hand corner lies on the curve of the graph). Because the function is increasing over the interval, all of the rectangles cover the entire region under the graph contained within their subinterval, plus a small portion of the space above the graph. The approximation will therefore be an overestimate of the actual area.

With both of these approximations, increasing the number of subintervals will give us a more accurate result. When our approximation produces an over estimate of the area under the graph, we call it an upper Riemann sum. When an under estimate is produced, we call it a lower Riemann sum. If we keep increasing the number of subintervals used, the upper and lower Riemann sums get closer and closer together, as they approach the same limit. This limit is of course the actual area under the curve, sometimes called the Riemann integral. We will use the term definite integral.

The concept of integration is based on the idea that, if we had an infinite number of rectangles, each of infinitesimal width, then adding together the areas of those rectangles would give us the exact area under the graph.

We can express this idea quite concisely using the following mathematical notation:

$$\int_a^b f\left(x\right) dx.$$

In case you are unfamiliar with the notation used here, we'll briefly explain what it means. The strange-looking symbol at the beginning is the integration symbol. It is based on the long 's' character, which is often mistaken for an 'f', and which can often be found in very old texts (for example, in the United States Bill of Rights, where it appears in the word "Congress"). This letter was chosen by the German mathematician Gottfried Wilhelm Leibniz to represent integration, because he thought of the integral as an infinite sum.

The subscripted and superscripted characters a and b immediately to the right of the integration symbol represent the lower and upper bounds respectively of the interval over which the function is defined, and over which we are integrating. This interval is also known as the domain of integration, with a as the lower limit of integration and b as the upper limit of integration. Next, we see the expression to be integrated - in this case the function $f(x)$ - which we call the integrand (the integrand is usually, though not always, a function). Last but not least we have the term dx, which tells us that we are integrating over x, which we refer to as the variable of integration. It can also be seen as representing an infinitesimally small increment of x.

Let's take things a step further. Suppose we have a continuous function $f(x)$ defined on the interval [a, b]. We will divide the interval into n subintervals of equal width. We'll call the width of each subinterval Δx, and we will use the letter i as an index when referring to a particular sub interval. Within each subinterval, we choose a distinguished point on the x axis, which we'll label x_i^*. The value of $f(x_i^*)$ will give us the height of the rectangle for each subinterval. The area of the rectangle within a given subinterval will therefore be:

$f(x_i^*) \, \Delta x.$

The Riemann sum (which we get by adding together the areas of the rectangles in all of the subintervals) will thus be:

$$S = \sum_{i=1}^{n} f\left(x_i^*\right)\Delta x.$$

However, the ultimate goal of integration is to get away from this idea of adding together the areas of a finite number of rectangles. Instead, we should be thinking about the sum of an infinite number of rectangles of infinitesimal width. To that end, we can express the definite integral of $f(x)$ over the interval [a, b] as:

$$\int_a^b f\left(x\right)dx = \lim_{n\to\infty}\sum_{i=1}^{n} f\left(x_i^*\right)\Delta x.$$

The definite integral (or Riemann interval) is thus the limit of the sum of $f(x_i^*)\,\Delta x$ for i = 1 to n, as n tends to infinity and Δx tends to zero. Note that the question of choosing a distinguished point within each interval at which to evaluate $f(x)$ becomes somewhat redundant, since Δx tends to zero anyway. All of which is all very interesting, of course, but we still need to be able to find the indefinite integral without having to calculate, and sum together, an infinite number of terms! Let's have a look at how we might do this. We'll start off by looking at a somewhat trivial example. The illustration below shows the graph of the function $f(x) = x$ for $0 \leq x \leq 1$.

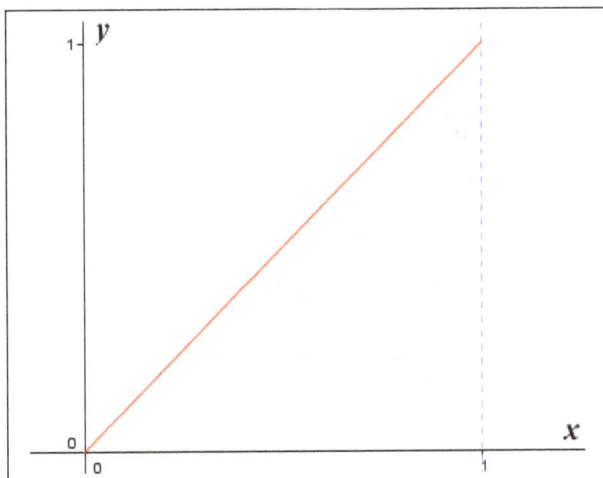

The graph of the function $f(x) = x$ for $0 \leq x \leq 1$.

Obviously, we can easily find the area under the graph in this case, since it will simply be the area of a right-angled triangle in which both legs are one unit in length. We are going to ignore that minor detail, however, and instead apply what we have learnt so far about the definite integral. Let's suppose that we are going to subdivide the interval [0, 1] into n subintervals. The width of each subinterval will therefore be given by:

$$\Delta x = \frac{1-0}{n} = \frac{1}{n}.$$

Although the choice of distinguished point within each subinterval is essentially unimportant, let's assume for the sake of argument that we are calculating a right-handed Riemann sum. The values of x taken by $f(x)$ in each subinterval will therefore be:

$$x_i^* = i\Delta x = i\frac{1}{n} = \frac{i}{n}$$

We can now formulate our Riemann sum:

$$S = \sum_{i=1}^{n} f(x_i)\Delta x = \sum_{i=1}^{n} \frac{i}{n}\cdot\frac{1}{n} = \frac{1}{n^2}\sum_{i=1}^{n} i$$

The last term in our Riemann sum is now simply the sum of consecutive integers, for which there is a well-known formula:

$$\sum_{i=1}^{n} i = \frac{n(n+1)}{2}$$

If we substitute this expression into our Riemann sum, we get:

$$S = \frac{1}{n^2}\cdot\frac{n(n+1)}{2} = \frac{n+1}{2n}$$

So the definite integral of $f(x) = x$ for $0 \le x \le 1$, if we take the limit as n tends to infinity, is:

$$\int_0^1 f(x)\,dx = \lim_{n\to\infty}\frac{n+1}{2n} = \frac{1}{2}$$

We could of course have arrived at the same answer by calculating the area of the triangular region under the graph (base multiplied by height, all over two), but we wanted to demonstrate a point, which is that the definite integral can be calculated algebraically as the limit of a Riemann sum. In order to persuade ourselves that this will work for non-linear functions as well, let's look at another example. This time we'll find the area under the graph of the function $f(x) = x^2$ for $0 \le x \le 2$. The graph is shown below:

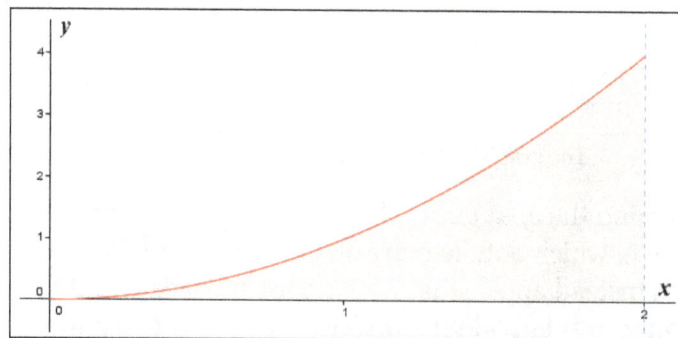

The graph of the function $f(x) = x^2$ for $0 \le x \le 2$.

As before, we are going to subdivide the interval [0, 2] into n subintervals, so the width of each subinterval is given by:

$$\Delta = \frac{2-0}{n} = \frac{2}{n}$$

Once again, although the choice of distinguished point within each subinterval is unimportant, let's assume that we are calculating a right-handed Riemann sum. The values of x taken by $f(x)$ in each subinterval will therefore be:

$$x_i^* = i\Delta x = i\frac{2}{n} = \frac{2i}{n}$$

Our Riemann sum will be as follows:

$$S = \sum_{i=1}^{n} f(x_i)\Delta x = \sum_{i=1}^{n}\left(\frac{2i}{n}\right)2\cdot\frac{2}{n} = \frac{8}{n^3}\sum_{i=1}^{n}i^2$$

The last term in our Riemann sum is now the sum of the squares of consecutive integers. Just as there is a well-known formula for the sum of consecutive integers, there is also a well-known formula for the sum of the squares of consecutive integers:

$$\sum_{i=1}^{n}i^2 = \frac{n(n+1)(2n+1)}{6}$$

If we substitute this expression into our Riemann sum, we get:

$$S = \frac{8}{n^3}\cdot\frac{n(n+1)(2n+1)}{6} = \frac{8n^2+12n+4}{3n^2}$$

So the definite integral of $f(x)$ for $0 \le x \le 2$, if we take the limit as n tends to infinity, is:

$$\int_0^2 f(x)\,dx = \lim_{n\to\infty}\frac{8n^2+12n+4}{3n^2} = \frac{8}{3}$$

This is the correct answer, and we have now demonstrated that it is possible to calculate the area under the graph of a non-linear function algebraically. In case you still need convincing, let's try one more example. This time we'll find the area under the graph of the function $f(x) = x^3 + 1$ for $0 \le x \le 1$. The graph is shown below:

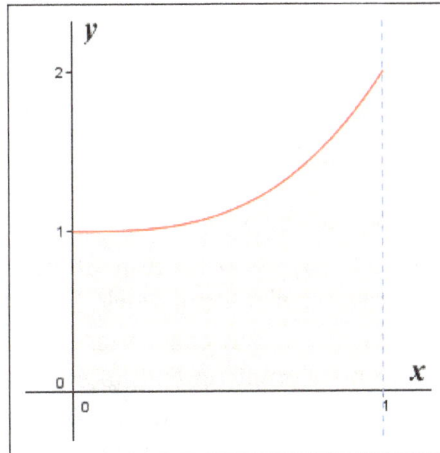

The graph of the function $f(x) = x^3 + 1$ for $0 \le x \le 1$.

As with the previous examples, we'll subdivide the interval [0, 1] into n subintervals, so the width of each subinterval is given by:

$$\Delta x = \frac{1-0}{n} = \frac{1}{n}$$

Although we have pretty much established that the choice of distinguished point within each sub-interval is unimportant, we will assume that we are calculating a right-handed Riemann sum. The values of x taken by $f(x)$ in each subinterval will therefore be:

$$x_i^* = i\Delta x = i\frac{1}{n} = \frac{i}{n}$$

Our Riemann sum is formulated as follows:

$$S = \sum_{i=1}^{n} f(x_i)\Delta x = \sum_{i=1}^{n}\left(\left(\frac{i}{n}\right)3+1\right).\frac{1}{n} = \frac{1}{n^4}\sum_{i=1}^{n}i^3$$

We now have a term in the Riemann sum that is the sum of consecutive integers cubed (i.e. raised to the power of three). Fortunately for us, there is also a (maybe not quite so well-known) formula for the sum of consecutive powers of three:

$$\sum_{i=1}^{n}i^3 = \frac{n^2(n+1)^2}{4}$$

If we substitute this expression into our Riemann sum, we get:

$$S = \frac{1}{n^4}.\frac{n^2(n+1)^2}{4}+1 = \frac{(n+1)^2}{4n^2}+1$$

So the definite integral of $f(x) = x^3 + 1$ for $0 \le x \le 1$, if we take the limit as n tends to infinity, are:

$$\int_0^1 f(x)\,dx = \lim_{n\to\infty}\frac{4n^2+(n+1)^2}{4n^2} = \frac{5}{4}$$

Once again, we can confirm that this is the correct answer. We now know that it is at least possible to calculate the area under the graph of a non-linear function algebraically, but the way we have been doing it seems like an awful lot of work. In fact, we have been fortunate with the examples shown in the sense that, in each case, there was a handy summation formula available that we could plug in to our calculation to make life easier. This will not always be the case - most of the time we need to work a lot harder to solve integral calculus problems algebraically.

Properties of Definite Integral

Property: The definite integral of the sum of two functions is equal to the sum of the definite integrals of these functions:

$$\int_a^b \left[f(x) + g(x) \right] dx = \int_a^b f(x) dx + \int_a^b g(x) dx.$$

Proof: By the definition of the definite integral,

$$I = \int_a^b \left[f(x) + g(x) \right] dx = \lim_{\lambda \to 0} \sum_{k=1}^n \left[f(\xi_k) + f(\xi_k) \right] \Delta x_k$$

Removing the square brackets under the sum and taking into account that the sum does not depend on the order of summation, we obtain:

$$I = \lim_{\lambda \to 0} \left[\sum_{k=1}^n f(\xi_k) \Delta x_k + \sum_{k=1}^n g(\xi_k) + g(\xi_k) \Delta x_k \right]$$

The limit of the sum equals to the sum of the limits, i.e.

$$I = \lim_{\lambda \to 0} \sum_{k=1}^n f(\xi_k) \Delta x_k + \lim_{\lambda \to 0} \sum_{k=1}^n g(\xi_k) \Delta x_k$$

By the definition of the definite integral these limits are the definite integrals on the right side.

Property: The constant coefficient c can be factored out:

$$\int_a^b cf(x) - c \int_a^b f(x) dx$$

The proof is similar to the proof of the first property.

Conclusion: The definite integral of the difference of two functions equals to the difference of the definite integrals of those functions:

$$\int_a^b \left[f(x) - g(x) \right] dx \int_a^b f(x) dx - \int_a^b g(x) dx$$

Proof: It follows from the first and the second property. Writing $f(x) - g(x) = f(x) + (-1)g(x)$ gives,

$$\int_a^b \left[f(x) - g(x) \right] dx = \int_a^b \left[f(x) + (-1) g(x) \right] dx = \int_a^b f(x) dx + (-1) \int_a^b g(x) dx$$

which is we wanted to prove.

Property: If $f(x) \geq 0$ for any $x \in [a; b]$, then,

$$\int_a^b f(x)\,dx \geq 0$$

Proof: If $f(x) \geq 0$ on $[a; b]$, then $f(x) \geq 0$ on any subinterval $[x_{k-1}; x_k]$, $k = 1, 2,\ldots , n$. Thus, for $\xi_k \in [xk_{-1}; x_k]$ also $f(\xi_k) \geq 0$. Multiplying the last inequality by the length of the kth subinterval gives $f(\xi_k)\, \Delta x_k \geq 0$, $k = 1, 2,\ldots , n$.

Adding n nonnegative quantities, we obtain nonnegative quantity:

$$\sum_{k=1}^n f(\xi_k)\Delta x_k \geq 0$$

By the limit theorem the limit of the nonnegative quantity as $\lambda \to 0$ is nonnegative, which proves the property.

If $f(x) \leq g(x)$ for any $x \in [a; b]$, then:

$$\int_a^b f(x)\,dx \leq \int_a^b g(x)\,dx$$

Proof: By assumption $g(x) - f(x) \geq 0$. Hence, by the

Property, $\int_a^b \left[g(x) - f(x)\right]dx \geq 0$

$$\int_a^b g(x)dx - \int_a^b f(x)\,dx \geq 0$$

which proves the statement.

Property: The absolute value of the definite integral of the function $f(x)$ is less than, or equal to, the definite integral of the absolute value of this function,

$$\left|\int_a^b f(x)\,dx\right| \leq \int_a^b |f(x)|\,dx$$

Proof: Here we use the property of the absolute value of the sum $|a + b| \leq |a| + |b|$ for n addends. By the definition of the definite integral,

$$\left|\int_a^b f(x)\,dx\right| = \left|\lim_{\lambda \to 0} f(\xi_k)\Delta x_k\right| = \lim_{\lambda \to 0}\left|\sum_{k=1}^n f(\xi_k)\Delta x_k\right| \leq$$

$$\leq \lim_{\lambda \to 0}\sum_{k=1}^n |f(\xi_k)\Delta x_k| = \lim_{\lambda \to 0}\sum_{k=1}^n |f(\xi_k)|\Delta x_k = \int_a^b |f(x)|\,dx$$

Property: If we change the limits of integration, then the sign of the integral changes,

$$\int\limits_b^a f(x)\,dx - \int\limits_a^b f(x)\,dx$$

Proof: If we define the definite integral $\int\limits_b^a f(x)\,dx,$ then the start point is b. If we assume that the definite integrals in this property exist, the limit does not depend on the partition. So in both definitions we can use the same partition. As well we can use in both definitions the same arbitrarily chosen points $\xi_1, \xi_2,\dots, \xi_{k-1}, \xi_k,\dots, \xi_n,$ because the limit does not depend on the choice of these points. If we move in direction from b to a, then the first partition point is $x_n = b$, next $x_{n-1},\dots, x_k,$ $x_{k-1},\dots, x_0 = a$. The start point of the kth subinterval x_k and the endpoint x_{k-1}. By the definition of the definite integral:

$$\int\limits_a^b f(x)\,dx = \lim_{\lambda\to 0}\sum_{k=n}^1 f(\xi_k)(x_{k-1} - x_k)$$

The integral sum in this definition is:

$$\sum_{k=n}^1 f(\xi_k)(x_{k-1} - x_k) = \sum_{k=n}^1 f(\xi_k)(-\Delta x_k) = -\sum_{k=n}^1 f(\xi_k)\Delta x_k$$

and, because the sum does not depend on the order of addition:

$$= \sum_{k=n}^1 f(\xi_k)(x_{k-1} - x_k) = -\sum_{k=1}^n f(\xi_k)\Delta x_k$$

The limit of the left side of this equality as $\lambda\to 0$ is $\int\limits_b^a f(x)$ and the limit of the right side is $-\int\limits_a^b f(x)\,dx.$

If the lower and upper limit of the definite integral is equal, then the definite integral equals to zero:

$$\int\limits_a^a f(x)\,dx = 0$$

Proof: Changing the limits of integration, we have by Property,

$$\int\limits_a^a f(x)\,dx = -\int\limits_a^a f(x)\,dx$$

or

$$2\int\limits_a^a f(x)\,dx = 0$$

which yields the assertion.

Property: Additivity property of the definite integral,

$$\int_a^b f(x)\,dx = \int_a^c f(x)\,dx + \int_c^b f(x)\,dx$$

Proof: First we assume that c is in the interval $[a; b]$, i.e. $a < c < b$. Defining the integral on the left side of this equality, we choose an arbitrary partition of the interval $[a; b]$, so that the first partition point is c. The further arbitrary partition of $[a; b]$ produces an arbitrary partition of the intervals $[a; c]$ and $[c; b]$. Thus, the integral sum for the whole interval $[a; b]$ can be written as the sum of the two integral sums,

$$\sum_{[a;b]} f(\xi_k)\Delta x_k = \sum_{[a;c]} f(\xi_k)\Delta x_k + \sum_{[c;b]} f(\xi_k)\Delta x_k$$

If the greatest length of the subintervals of $[a; b]$ $\lambda \to 0$, then the greatest lengths of the subintervals of $[a; c]$ and $[c; b]$ approach zero as well. Therefore, taking the limits as $\lambda \to 0$ on both sides completes the proof.

If c is outside the interval [a; b], suppose $c > b > a$, then,

$$\int_a^c f(x)\,dx = \int_a^b f(x)\,dx + \int_b^c f(x)\,dx$$

It follows,

$$\int_a^b f(x)\,dx = \int_a^c f(x)\,dx - \int_b^c f(x)\,dx$$

and changing the limits we have,

$$\int_a^b f(x)\,dx = \int_a^c f(x)\,dx + \int_c^b f(x)\,dx$$

In similar way we can prove that this property holds if c < a.

Property: If m is the least value of $f(x)$ and M is the greatest value of $f(x)$ on the interval $[a; b]$, then,

$$m(b-a) \le \int_a^b f(x)\,dx \le M(b-a)$$

Proof: The proofs of these two inequalities are similar and we prove only the right hand inequality.

As assumed, the greatest value of the function $f(x)$ on $[a; b]$ is M. Thus, $f(\xi_k) \le M$ for any arbitrarily chosen $\xi_k \in [x_{k-1}; x_k]$ for each $k = 1, 2,\dots, n$. Multiplying this inequality by Δx_k gives,

$$f(\xi_k)\Delta x_k \le M\Delta x_k$$

Adding these products, we obtain:

$$\sum_{k=1}^{n} f(\xi_k)\Delta x_k \le \sum_{k=1}^{n} M\Delta x_k$$

$$= M(x_1 - x_0 + x_2 - x_1 + x_3 - x_2 + \ldots + x_n - x_{n-1}) = M(b - a),$$

because $x_0 = a$ and $x_n = b$.

There is constant on the right side of the inequality,

$$\sum_{k=1}^{n} f(\xi_k)\Delta x_k \le M(b-a)$$

and taking the limit on both sides of this inequality as $\lambda \to 0$ gives the assertion.

Property: If the function $f(x)$ is continuous on $[a; b]$, then there exists at least one point $\xi \in [a; b]$ such that:

$$\int_a^b f(x)dx = f(\xi)(b-a)$$

Proof: The function continuous in the closed interval has the least value m and the greatest value M on this interval, hence, there holds the Property. Dividing the both sides of the both inequalities by the length of the interval of integration $b - a$ gives,

$$m \le \frac{1}{b-a}\int_a^b f(x)dx \le M$$

Consequently,

$$\frac{1}{b-a}\int_a^b f(x)dx$$

is between the least and the greatest value. The function continuous on $[a; b]$ has any value between the least and the greatest. Therefore, there exists at least one point $\xi \in [a; b]$, where the function obtains this value, that is,

$$f(\xi) = \frac{1}{b-a}\int_a^b f(x)dx.$$

The multiplication of both sides of this equality by $b-a$ completes the proof. The value $f(\xi)$ is called the *mean value* of the function $f(x)$ on the interval $[a; b]$.

Computation of Definite Integral: Newton-Leibnitz Formula

Suppose $f(x)$ is defined on $[a; b]$. Let us define on $[a; b]$ the function of the upper limit of the definite integral,

$$\Phi(x) = \int_a^x f(t)\,dt$$

Theorem: If the function $f(x)$ is continuous on $[a; b]$, then $\Phi(x)$ is differentiable on $(a; b)$ and $\Phi'(x) = f(x)$.

Proof: We use the definition of the derivative of $\Phi(x)$,

$$\Phi'(x) = \lim_{\Delta x \to 0} \frac{\Phi(x + \Delta x) - \Phi(x)}{\Delta x}$$

By additivity property of the definite integral:

$$\Phi(x + \Delta x) - \Phi(x) = \int_a^{x+\Delta x} f(t)\,dt - \int_a^x d(t)\,dt =$$

$$\int_a^b f(t)\,dt + \int_a^{x+\Delta x} f(t)\,dt - \int_a^x f(t)\,dt = \int_a^{x+\Delta x} f(t)\,dt.$$

As assumed, the function $f(x)$ is continuous on $[a; b]$. Hence, by the mean value property, there exists $\xi \in [x; x + \Delta x]$ such that:

$$\Phi(x + \Delta x) - \Phi(x) = f(\xi)(x + \Delta x - x) = f(\xi)\Delta x$$

Consequently:

$$\frac{\Phi(x + \Delta x) - \Phi(x)}{\Delta x} = f(\xi)$$

In the definition of the derivative $\Delta x \to 0$. It follows that $x + \Delta x \to x$ and since ξ is a point between x and $x + \Delta x$, then $\xi \to x$ also. Thus:

$$\Phi'(x) = \lim_{\Delta x \to 0} f(\xi) = \lim_{\xi \to x} f(\xi)$$

and the third condition of continuity of $f(x)$ gives $\Phi'(x) = f(x)$, which is we wanted to prove.

By theorem, the function $\Phi(x)$ is an antiderivative of $f(x)$. If $F(x)$ is the known antiderivative of $f(x)$ (by the table of integrals or by some technique of integration), then by Corollary antiderivatives $\Phi(x)$ and $F(x)$ differ at most by a constant, i.e. $\Phi(x) = F(x) + C$. According to the definition of $\Phi(x)$,

$$F(x) + C = \int_a^x f(t)\,dt.$$

Taking in this equality x = a, we obtain by:

$$F(a)+C=\int_a^a f(t)dt=0$$

which yields C = −F(a). Substituting C $F(x)+C=\int_a^x f(t)dt.$

$$F(x)-F(a)=\int_a^x f(t)dt$$

and taking in the last equality x = b, we obtain:

$$F(b)-F(a)=\int_a^b f(t)dt.$$

Consequently, the antiderivative familiar from the indefinite integral is the appropriate tool to evaluate the definite integral. Now we take in $F(b)-F(a)=\int_a^b f(t)dt.$ for the variable of integration x again. To facilitate the computation we use the notation:

$$F(b)-F(a)=F(x)\big|_a^b$$

Finally, we formulate the result obtained as a theorem.

Theorem: If the function f(x) is continuous on [a; b] and F(x) is the antiderivative of f(x), then:

$$\int_a^b f(x)dx = F(x)\big|_a^b = F(b)-F(a),$$

Example:

$$\int_a^e \frac{dx}{x} = In\,x\big|_1^e = In\,e - In\,1 = 1$$

$$\int_0^1 \frac{xdx}{\sqrt{1+x^2}}$$

For the integration we use the equality d(1 + x²) = 2xdx and find:

$$\int_0^1 \frac{xdx}{\sqrt{1+x^2}} = \frac{1}{2}\int_0^1 \frac{2xdx}{\sqrt{1+x^2}} = \frac{1}{2}\int_0^1 \frac{d(1+x^2)}{\sqrt{1+x^2}}$$

$$= \frac{1}{2}.2\sqrt{1+x^2}\,\big|_0^1 = \sqrt{2}-1.$$

Example: Compute the mean value of the function $f(x) = x^2$ on $[1; 3]$. By the mean value formula:

$f(\xi) = \dfrac{1}{b-a}\displaystyle\int_a^b f(x)\,dx.$ we find,

$$\frac{1}{3-1}\int_1^3 x^2\,dx = \frac{1}{2}\frac{x^3}{3}\bigg|_1^3 = \frac{1}{2}\left(\frac{27}{3}-\frac{1}{3}\right) = \frac{13}{3} = 4\frac{1}{3}$$

Change of Variable in Definite Integral

The choice of the new variable depends on the function to be integrated. These principals are familiar from the indefinite integral.

If we compute the definite integral, we are interested in its value, not in the antiderivative of the initial function. This is because after the integration by change of variable in the definite integral we don't re-substitute the initial variable. Instead of it we compute the limits of integration for the new variable.

Changing the variable x = φ (t) in the definite integral,

$$\int_a^b f(x)\,dx$$

we find $dx = \varphi'(t)dt$. The equation $\varphi(t) = a$ gives the lower limit for the new variable $t = \alpha$ and the equation $\varphi(t) = b$ gives the upper limit $t = \beta$. The change of variable formula is,

$$\int_a^b f(x)\,dx = \int_a^\beta f\left[(\varphi(t))\right]\varphi'(t)\,dt$$

Example: Compute $I = \displaystyle\int_0^2 \sqrt{8-x^2}\,dx$. To remove the irrationality we change the variable $x = 2\sqrt{2}\sin t$.

Then $dx = 2\sqrt{2}\cos t\,dt$

$$\sqrt{8-x^2} = \sqrt{8-8\sin^2 t} = \sqrt{8\cos^2 t} = 2\sqrt{2}\cos t$$

We determine the limits for the new variable t. If x = 0, then sin t = 0, it follows t = 0. If x = 2, then

$2\sqrt{2}\sin t = 2$ or $\sin t = \dfrac{\sqrt{2}}{2}$, hence, $t = \dfrac{\pi}{4}$.

Thus,

$$I = \int_a^{\frac{\pi}{4}} 2\sqrt{2}\cos t.2\sqrt{2}\cos t\,dt = 8\int_a^{\frac{\pi}{4}} \cos^2 t\,dt =$$

$$= 4\int_a^{\frac{\pi}{4}}(1+\cos 2t)\,dt = 4\int_a^{\frac{\pi}{4}} dt + 2\int_a^{\frac{\pi}{4}}\cos 2t\,d(2t) =$$

$$= 4t\big|_0^{\frac{\pi}{4}} + \sin 2t\big|_0^{\frac{\pi}{4}} = \pi + 2.$$

Improper Integral over Infinite Interval

Both of these are integrals that are called improper integrals. In the first kind of improper integrals one or both of the limits of integration are infinity.

Let the function $f(x)$ be defined and continuous on the infinite interval $[a;\infty)$. If for any $N \in [a;\infty)$ there exists the definite integral $\int_a^N f(x)\,dx$ and there exists the limit $\lim\limits_{N\to\infty}\int_a^N f(x)\,dx$, then this limit is called the improper integral with the infinite upper limit and denoted $\int_a^\infty f(x)\,dx$.

$$\int_a^\infty f(x)\,dx = \lim_{N\to\infty}\int_a^N f(x)\,dx$$

If the limit exists and is a finite number, then the improper integral is said to be convergent. If the limit does not exist or the limit is infinite, then the improper integral is said to be divergent.

Thus, to compute the improper integral, we first have to compute the definite integral over $[a\,;N]$ and next find the limit of this result as $N\to\infty$.

Example: Evaluate $\int_0^\infty \dfrac{dx}{1+x^2}$,

$$\int_a^\infty \frac{dx}{1+x^2} = \lim_{N\to\infty}\int_0^N \frac{dx}{1+x^2} = \lim_{N\to\infty}\text{archan } x\Big|_0^N = \lim_{N\to\infty}(\arctan N - \text{archtan } 0) = \frac{\pi}{2}$$

So, this improper integral is convergent.

Let the function $f(x)$ be defined and continuous on the infinite integral $(-\infty;b]$. If for any $M \in (-\infty;b]$ there exists $\int_M^b f(x)\,dx$ and there exists the limit $\lim\limits_{M\to-\infty}\int_M^b f(x)\,dx$, then this limit is called the im-proper integral with the infinite lower limit and denoted $\int_{-\infty}^b f(x)\,dx$.

$$\int_{-\infty}^b f(x)\,dx = \lim_{M\to-\infty}\int_M^b f(x)\,dx$$

The convergence and divergence of this improper integral are defined in the same way as in the previous case.

If the function $f(x)$ is defined and continuous in $(-\infty;\infty)$, then the improper integral over $(-\infty;\infty)$ is defined as,

$$\int_{-\infty}^\infty f(x)\,dx = \int_{-\infty}^c f(x)\,dx + \int_c^\infty f(x)\,dx$$

where c is any finite real number.

If both of the improper integrals on the right side of this equality are convergent, then this improper integral is said to be convergent. If at least one of the improper integrals on the right side of this equality is divergent, then this improper integral is said to be divergent.

Example: Let a > 0 and let us decide for which values of α the improper integral $\int\limits_a^b \dfrac{dx}{x^\alpha}$ is convergent

and for which values of α it is divergent. Denote this improper integral by I and find,

$$I = \int\limits_a^\infty \frac{dx}{x^a} = \lim_{N\to\infty} \int\limits_a^N \frac{dx}{x^a}$$

If $\alpha \neq 1$, then:

$$I = \lim_{N\to\infty} \frac{x^{-\alpha+1}}{-\alpha+1}\bigg|_a^N = \lim_{N\to\infty}\left(\frac{N^{-\alpha+1}}{-\alpha+1} - \frac{a^{-\alpha+1}}{-\alpha+1} \right)$$

If $\alpha > 1$, then:

$$\lim_{N\to\infty}\left(\frac{1}{(1-\alpha)N^{\alpha-1}} - \frac{1}{(1-\alpha)a^{\alpha-1}} \right) = \frac{1}{(\alpha-1)a^{\alpha-1}}$$

that means, the improper integral is convergent. If $\alpha < 1$, then:

$$\lim_{N\to\infty}\left(\frac{N^{1-\alpha}}{(1-\alpha)} - \frac{a^{1-\alpha}}{1-\alpha} \right) = \infty$$

That is, the improper integral is divergent.

If $\alpha = 1$, then:

$$\int\limits_a^\infty \frac{dx}{x} = \lim_{N\to\infty} \ln x \Big|_a^N = \lim_{N\to\infty}\left(\ln N - \ln a \right) = \infty$$

thus, the improper integral is divergent again.

Consequently, the improper integral $\int\limits_a^\infty \dfrac{dx}{x^\alpha}$ is convergent, if $\alpha > 1$ and divergent, if $\alpha \leq 1$.

In many cases we are rather interested in the convergence of the improper integral than in the actual value of this integral. Moreover, sometimes an improper integral is too difficult to evaluate, but we still need to know, is it convergent or not. One technique is to compare it with a known integral. The theorems below, called the comparison theorems, enable us to decide whether the improper integral is convergent or divergent. We formulate these theorems for the improper integral with infinite upper limit. These theorems hold as well for the improper integrals with infinite lower limit and in case, if both limits are infinite.

We assume, that we know whether the improper integral $\int\limits_a^\infty \varphi(x)\,dx$ is convergent or divergent.

Theorem: Suppose that $f(x)$ and $\varphi(x)$ are two continuous on $[a;\infty)$ functions such that $0 \leq f(x) \leq \varphi(x)$ on this interval. Then the convergence of the improper integral,

$$\int\limits_a^\infty \varphi(x)\,dx$$

yields the convergence of the improper integral,

$$\int\limits_a^\infty f(x)\,dx.$$

Suppose that $f(x)$ and $\varphi(x)$ are two continuous on $[a;\infty)$ functions such that $0 \leq \varphi(x) \leq f(x)$ on this interval. Then the divergence of the improper integral $\int\limits_a^\infty \varphi(x)\,dx$ yields the divergence of the improper integral $\int\limits_a^\infty f(x)\,dx$.

Theorem: Suppose that two continuous on $[a;\infty)$ functions $f(x)$ and $\varphi(x)$ are equivalent in the limiting process $x \to \infty$.

The improper integral is called absolutely convergent, if the improper integral,

$$\int\limits_a^\infty |f(x)|\,dx$$

is convergent.

Theorem: The absolute convergence of yields the convergence of this improper integral.

Example: Decide on the convergence or divergence of,

$$\int\limits_a^\infty \frac{\arctan x\,dx}{1+x^2}$$

In the half-interval $[0;\infty)$ there holds $\arctan x \leq \dfrac{\pi}{2}$.

By example, the improper integral $\int\limits_1^\infty \dfrac{dx}{1+x^2}$ is convergent. Applying theorem for $f(x) = \dfrac{\arctan x}{1+x^2}$ and $\varphi(x) = \dfrac{\pi}{2} \cdot \dfrac{1}{1+x^2}$, we conclude that the given improper integral is convergent.

Example: Decide on the convergence or divergence of $\int\limits_2^\infty \dfrac{dx}{x-1}$.

In the limiting process x → ∞, the functions $f(x) = \dfrac{1}{x-1}$ and $\varphi(x) = \dfrac{1}{x}$ are equivalent because:

$$\lim_{x\to\infty} \frac{\frac{1}{x-1}}{\frac{1}{x}} = \lim_{x\to\infty} \frac{x}{x-1} = 1$$

By example the improper integral $\int_2^\infty \dfrac{dx}{x}$ is divergent. Thus, by theorem, the given improper integral is also divergent.

Example: Decide on the convergence or divergence of $\int_1^\infty \dfrac{\sin x\, dr}{x^2}$

For any x ∈ ℝ there holds $\left|\dfrac{\sin x}{x^2}\right| \le \dfrac{1}{x^2}$. By example the improper integral $\int_2^\infty \dfrac{dx}{x^2}$ is convergent. By theorem this improper integral is absolutely convergent.

Improper Integrals of Unbounded Functions

Suppose that the function $f(x)$ is unbounded in a neighborhood of the right endpoint b of the interval $[a; b]$.

If for any $\varepsilon > 0$ there exists the definite integral $\int_a^{b-\varepsilon} f(x)dx$ and there exists the limit $\lim_{\varepsilon\to0} \int_a^{b-\varepsilon} f(x)dx$ then this limit is called the improper integral of the unbounded function at the upper limit and denoted $\int_a^b f(x)dx$.

We evaluate the improper integral of the unbounded function in the neighborhood of the upper limit b, using the formula,

$$\int_a^b f(x)dx = \lim_{\varepsilon\to0} \int_a^{b-\varepsilon} f(x)dx$$

Improper integrals are often written symbolically just like standard definite integrals.

Suppose that the function $f(x)$ is unbounded in a neighborhood of the left endpoint a of the interval $[a; b]$.

If for any $\varepsilon > 0$ there exists the definite integral $\int_{a+\varepsilon}^b f(x)dx$ and there exists the limit $\lim_{\varepsilon\to0} \int_{a+\varepsilon}^b f(x)dx$

,then this limit is called the improper integral of the unbounded function at the lower limit and denoted $\int_a^b f(x)dx$.

By definition the improper integral of the unbounded function at the lower limit a we evaluate by the formula,

$$\int_a^b f(x)dx = \lim_{\varepsilon \to 0} \int_{a+\varepsilon}^b f(x)dx$$

If the function f(x) is unbounded in some interior point c of [a; b], then we use the additivity property on the integral and write,

$$\int_a^b f(x)dx = \int_a^c f(x)dx \int_c^b f(x)dx$$

and evaluate the first addend.

If the limits are finite, then the improper integral is said to be convergent. If these limits either does not exist or are infinite, then this improper integral is said to be divergent.

The improper integral of the unbounded function is said to be absolutely convergent if the improper integral,

$$\int_a^b |f(x)|dx$$

is convergent.

Example: Let us find how the convergence or divergence of,

$$\int_a^b \frac{dx}{(b-x)^\alpha}$$

depends on the exponent a.

The integrand $\dfrac{1}{(b-x)^\alpha}$ is unbounded in the neighborhood of the upper limit b:

$$\int_a^b \frac{dx}{(b-x)^\alpha} = \lim_{\varepsilon \to 0} \int_a^{b-\varepsilon} \frac{dx}{(b-x)^\alpha}$$

Suppose a ≠ 1. Using the equality of the differentials d(b – x) = – dx,

we obtain,

$$\lim_{\varepsilon \to 0} \int_a^{b-\varepsilon} \frac{dx}{(b-x)^\alpha} = -\lim_{\varepsilon \to 0} \int_a^{b-\varepsilon} (b-x)^{-\alpha} d(b-x) = -\lim_{\varepsilon \to 0} \frac{(b-x)^{-\alpha+1}}{-\alpha+1}\bigg|_a^{b-\varepsilon} =$$

$$= -\lim_{\varepsilon \to 0}\left[\frac{\varepsilon^{-\alpha+1}}{-\alpha+1} - \frac{(b-a)^{-\alpha+1}}{-\alpha+1}\right] = \lim_{\varepsilon \to 0}\left[\frac{(b-a)^{1-\alpha}}{1-\alpha} - \frac{\varepsilon^{1-\alpha}}{1-\alpha}\right]$$

If $\alpha > 1$, then $\alpha - 1 > 0$ and $\lim\limits_{\varepsilon \to 0} \varepsilon^{\alpha-1} = 0$, Hence,

$$\lim_{\varepsilon \to 0} \frac{\varepsilon^{\alpha-1}}{1-\alpha} = \lim_{\varepsilon \to 0} \frac{1}{(1-\alpha)\varepsilon^{\alpha-1}} = \infty$$

that means the improper integral is divergent.

If $\alpha < 1$, then $1 - \alpha > 0$ and $\lim\limits_{\varepsilon \to 0} \varepsilon^{\alpha-1} = 0$, thus,

$$\lim_{\varepsilon \to 0} \left[\frac{(b-a)^{1-\alpha}}{1-\alpha} - \frac{\varepsilon^{1-\alpha}}{1-\alpha} \right] = \frac{(b-a)^{1-\alpha}}{1-\alpha},$$

the improper integral is convergent.

If $\alpha = 1$, then:

$$\lim_{\varepsilon \to 0} \int_a^{b-\varepsilon} \frac{dx}{(b-x)^\alpha} = \lim_{\varepsilon \to 0} \int_a^{b-\varepsilon} \frac{d(b-x)}{b-x} = -\lim_{\varepsilon \to 0} In|b-x| \Big\|_a^{b-\varepsilon} =$$

$$= \lim_{\varepsilon \to 0} \left(In|b-a| - In|\varepsilon| \right) = \infty,$$

the improper integral is divergent.

Consequently, the improper integral is convergent if $\alpha < 1$ja and divergent if $\alpha \geq 1$.

For the improper integrals of unbounded functions there hold the analogous theorems as for the improper integrals over unbounded intervals.

Theorem: If the functions $f(x)$ and $\varphi(x)$ continuous in the half interval $[a; b)$ satisfy the condition $0 \leq f(x) \leq \varphi(x)$ then the convergence of the improper integral,

$$\int_a^b \varphi(x)\,dx$$

yields the convergence of the improper integral,

$$\int_a^b f(x)\,dx$$

Theorem: If the functions $f(x)$ and $\varphi(x)$ continuous in the half interval $[a; b)$ are equivalent in the limiting process $x \to b$ then the convergence. The absolute convergence of the improper integral yields the convergence of that integral.

Approximate Computation of Definite Integral

Applying the Newton-Leibnitz formula to evaluate the definite integral, we have to find the antiderivative of the integrand. A lot of quite a simple functions, for instance:

$$e^{-x2}, \frac{\sin x}{x} \text{ and } \frac{1}{In x}$$

Don't have antiderivative among elementary functions. Thus, the Newton-Leibnitz formula is not applicable. In this case we use the approximate formulas to evaluate the definite integral. One of those approximate formulas is called trapezoidal rule.

Let us have an integral $\int_a^b f(x)dx$ for a continuous function $f(x) \geq 0$. We divide the interval [a; b] into n subintervals of equal width. So we obtain a partition:

$$a = x_0, x_1, x_2, \ldots x_{k-1}, x_k, \ldots, x_n = b$$

Hence, $x_k - x_{k-1} = h$ for any $k = 1, 2, \ldots n$ and the dividing points are $x_0 = a$, $x_1 = a + h$, $x_2 = a + 2h, \ldots, x_k = a + k_h, \ldots, x_n = a + n_h = b$.

The vertical lines $x = x_k$, $k = 1, 2 \ldots n-1$ divide the area ab BA under the graph into n areas $PQRS$. If we substitute the curve between R and S by the straight line RS, we obtain the trapezoid $PQRS$, whose parallel sides PS and QR have the lengths $f(x_{k-1})$ and $f(x_k)$, respectively. The length of one subdivision h is the height of trapezoid $PQRS$ and the area of this trapezoid is:

$$S_k = \frac{f(x_{k-1}) + f(x_k)}{2} \cdot h$$

The sum of the areas of n trapezoids PQRS equals approximately to the area under the graph abBA. If n is increasing, then the accuracy of this,

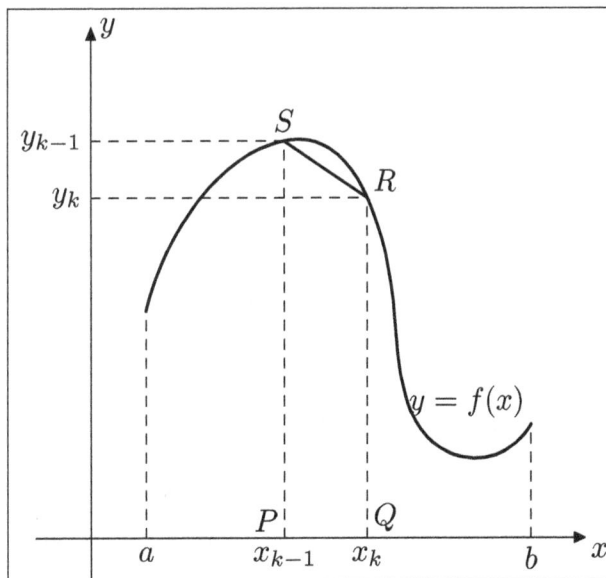

approximation becomes higher. The area under the graph is the value of the definite integral. Thus, the definite integral equals approximately to the sum of the areas of trapezoids PQRS:

$$\int_a^b f(x)dx \approx S_1 + S_2 + \ldots + S_n = \frac{f(x_0) + f(x_1)}{2} \cdot h + \frac{f(x_1) + f(x_2)}{2} \cdot h + \ldots + \frac{f(x_{n-1}) + f(x_n)}{2} \cdot h$$

Factoring out $\dfrac{h}{2}$, we have the approximate formula,

$$\int_a^b f(x)\,dx \approx \frac{h}{2}\left(f(x_0)+2f(x_1)+2f(x_2)+\ldots+2f(x_{n-1})+f(x_n)\right)$$

which is called trapezoidal rule, Notice that all the values of the function are multiplied by 2, except the values at the endpoints $y_0 = f(a)$ and $y_n = f(b)$.

Example: Compute by trapezoidal rule $\int_0^2 x^2\,dx$.

To compare the result with exact value, we calculate first this definite integral by Newton-Leibnitz formula $\int_0^2 x^2\,dx = \dfrac{x^3}{3}\Big|_0^2 = \dfrac{8}{3} = 2.$

Now we compute this definite integral by trapezoidal formula. First we divide the interval of integration into four four equal parts [0; 2], that means n = 4. The length of one subdivision is $h = \dfrac{2-0}{4} = 0,5$ and dividing points are $x_0 = 0$, $x_1 = 0,5$, $x_2 = 1$, $x_3 = 1,5$ and $x_4 = 2$.

Evaluating the function $f(x) = x^2$ at these points, we have $f(x_0) = 0$, $f(x_1) = 0,25$, $f(x_2) = 1$, $f(x_3) = 2,25$ and $f(x_4) = 4$:

$$\int_0^2 x^2\,dx \approx\sim 0,25\left(0+2\cdot0,25+2\cdot1+2\cdot2,25+4\right) = 2,75$$

Next we compute this integral by trapezoidal rule again, dividing the interval of integration [0; 2] into eight equal parts. Then the length of one subdivision is $h = 0,25$ and the dividing points are $x_0 = 0$, $x_1 = 0,25$, $x_2 = 0,5$, $x_3 = 0,75$, $x_4 = 1$, $x_5 = 1,25$, $x_6 = 1,5$, $x_7 = 1,75$ and $x_8 = 2$.

The values of the function $f(x) = x^2$ at these points are $f(x_0) = 0$, $f(x_1) = 0,0625$, $f(x_2) = 0,25$, $f(x_3) = 0,5625$, $f(x_4) = 1$, $f(x_5) = 1,5625$, $f(x_6) = 2,25$, $f(x_7) = 3,0625$ and $f(x_8) = 4$.

$$\int_0^2 x^2\,dx \approx 0,125\left(0+2\cdot0,0625+2\cdot0,5625+2\cdot1+2\cdot1,5625+2\cdot2,25+2\cdot3,0625+4\right) = 2,6875.$$

Techniques of Integration

Integration by Substitution

Integration by substitution, sometimes called u-substitution, is one such method. the chain rule gave us a formula that allowed us to differentiate composite functions. A composite function is a function in which one function (the outer function) is applied to the output of another function (the inner function). The chain rule works by allowing us to substitute a simple variable for the inner function so that we can differentiate the outer function first, without worrying about the inner function. The substitution rule performs a similar role for integration. It simplifies a composite function and makes it easier for us to integrate.

Let's start with a relatively straightforward example.

Consider the following integral:

$$\int (x+4)^5 \, dx$$

Before we proceed, consider for a moment how we would calculate the following integral:

$$\int x^5 \, dx$$

This is simply a matter of applying the power rule for integration, which is:

$$\int ax^n \, dx = a\frac{x^{n+1}}{n+1} + C$$

Remembering that in this case the constant coefficient a is one, we get:

$$\int x^5 \, dx = \frac{x^6}{6} + C$$

Applying power:

$$\int (x+4)^5 \, dx = \frac{(x+4)^6}{6} + C$$

The result is in fact correct, because we deliberately chose an example that would work, but it would be very dangerous to assume that we can use the power rule to solve this type of problem. Things are not as simple as they might seem. Let's see what happens when we use substitution here. In this case, it is relatively obvious that we need to substitute a variable for $x + 4$. The variable name used by convention is u. That's why integration by substitution is often called "u-substitution". The original integrand $(x + 4)^5$ now becomes u^5. However, this changes things, because the variable of integration is now u and not x. We therefore need to make a substitution for the term dx as well.

Remember that dx and du are both differentials. We can find du in terms of dx as follows:

$$du = \left(\frac{du}{dx}\right)dx$$

Differentiating x will always give us one. In this case, because u = x + 4, we can see that differentiating u will also give us one, so du will be equal to dx. Our substitutions will therefore give us the following:

$$\int (x+4)^5 \, dx = \int u^5 \, du = \frac{u^6}{6} + C$$

We can now substitute x + 4 back in for u to give us the following:

$$\int (x+4)^5 dx = \frac{(x+4)^6}{6} + C$$

Consider the following integral:

$$\int (2x+5)^3 dx$$

On the face of it, this problem is just like the one we solved in the previous example. If we apply the power rule, however, we get:

$$\int (2x+5)^3 dx = \frac{(2x+5)^4}{4} + C$$

Is this correct? Let's use u-substitution and see what happens. Once again, we'll substitute the variable u for the expression inside the brackets, i.e.

$$u = 2x + 5$$

Remember that because the variable of integration is now u and not x, we need to make a substitution for dx as well. Differentiating x will, as always, give us one. This time, though, because $u = 2x + 5$, differentiating u will give us two. In terms of dx, therefore, du is given by:

$$du = \left(\frac{du}{dx} \right) dx = 2dx$$

For this example, therefore, our substitutions will give us the following:

$$\int (2x+5)^3 dx = \int u^3 \frac{1}{2} du = \frac{u^4}{8} + C$$

We can now substitute 2x + 5 back in for u (we call this back substitution) to give us the following:

$$\int (2x+5)^3 dx = \frac{(2x+5)^4}{8} + C$$

This is, of course, not the same answer we got using the power rule, which demonstrates why we can't simply use the power rule to solve this type of problem.

The substitutions we make when attempting to solve integration problems often take the general form,

$$u = ax + b$$

where a is a constant coefficient, x is a variable, and b is a constant value. Let's look at another example where this kind of substitution is used.

Suppose we want to find the following integral:

$$\int \cos(5x+3)\,dx$$

The obvious substitution here will be u = 5x + 3. As before, we have:

$$du = \left(\frac{du}{dx}\right)dx$$

Differentiating x always gives us one, and differentiating 5x + 3 will give us five, so:

$$\frac{du}{dx} = 5$$

and

$$du = \left(\frac{du}{dx}\right)dx = 5dx$$

Although we deal with the integrals of trigonometric functions elsewhere in this section, it shouldn't come as too much of a surprise to learn that the integral of cos (x) is sin (x), since we have by now established that integration and differentiation are inverse operations, and from your work on differential calculus you may recall that the derivative of sin (x) is cos (x). The solution to this problem is therefore as follows:

$$\int \cos(5x+3)\,dx = \int \frac{1}{5}\cos(u)\,du = \frac{1}{5}\sin(u) + C$$

We can now substitute 5x + 3 back in for u to give us the following:

$$\int \cos(5x+3)\,dx = \frac{\sin(5x+3)}{5} + C$$

We can in fact generalise this result to find the integral of any expression in the form cos (ax + b). We simply use the general form of the substitution, i.e. u = ax + b, which gives us the following result:

$$\int \cos(ax+b)\,dx = \frac{1}{a}\int \cos(u)\,du$$

$$= \frac{1}{a}\sin(u) + C$$

$$= \frac{1}{a}\sin(ax+b) + C$$

Integrating the Product of Two Functions

As we have said previously, integration does not have a direct equivalent of the product rule we use to find the derivative of the product of two (or more) functions. Fortunately, there are a range of methods we can use to deal with problems of this sort, some of which will involve integration by substitution. Integration by substitution does for integration what the chain rule does for differentiation. It gives us a way to integrate composite functions. We can express the notion of integration by substitution somewhat more formally using the substitution rule:

$$\int f\left(g\left(x\right)\right)g'\left(x\right)dx = \int f\left(u\right)du$$

where,

$$u = g\left(x\right)$$

and,

$$du = g'\left(x\right)dx$$

Note that $g(x)$ is the function we are substituting u for, and $g'(x)$ is its derivative. What this means in essence is that if the expression we wish to integrate is in the form $f(g(x))\,g'(x)$, our integration task is going to be fairly straightforward. You may not quite see how that works yet, so let's look at an example. Consider the following integral:

$$\int \sin\left(x^2\right)2x\,dx$$

It should be fairly obvious that we will be substituting u for the inner function x^2. And, if we differentiate x^2, we get $2x$. Our integral now simply becomes:

$$\int \sin\left(u\right)du = -\cos\left(u\right) + C$$

And substituting back in for u we get:

$$\int \sin\left(x^2\right)2x\,dx = -\cos\left(x^2\right) + C$$

Of course, not all of the integral problems we come across are going to be in such a convenient format, as we have seen. In fact, it's reasonable to assume that most of them certainly won't be. Let's make life ever so slightly more complicated. Consider the following integral:

$$\int \sin\left(x^2\right)8x\,dx$$

This is almost the same as the previous problem, except that we now have 8x instead of 2x, which rather messes things up. Well, not really. All we really need to do is to rearrange things a little, like this:

$$4\int \sin\left(x^2\right)2x\,dx$$

We can do this because, as you may recall, the constant coefficient rule (or constant multiple rule) tells us that the indefinite integral of c. $f(x)$, where $f(x)$ is some function and c represents a constant coefficient, is equal to the indefinite integral of $f(x)$ multiplied by c. In this case, we have pulled four out as our constant coefficient, which leaves us with exactly the same integrand that we had in the previous example. We will once again substitute u for the inner function x^2, so our integral becomes:

$$4\int \sin(u)\,du = -4\cos(u) + C$$

And substituting back in for u we get:

$$\int \sin(x^2)8x\,dx = -4\cos(x^2) + C$$

Let's look at another example where we can do something similar. Consider the following integral:

$$\int x^4 \sin(x^5)\,dx$$

If we let $u = x^5$, then du will be $5x^4$. We have x^4 as part of the integral, which is not quite what we want. We can get around the problem as, however, by rearranging things a little:

$$\frac{1}{5}\int 5x^4 \sin(x^5)\,dx$$

We can now make our substitutions and perform the integration:

$$\frac{1}{5}\int \sin(u)\,du = -\frac{1}{5}\cos(u) + C$$

Substituting x^5 back in for u, we can now rewrite our integral as:

$$\int x^4 \sin(x^5)\,dx = -\frac{1}{5}\cos(x^5) + C$$

Let's look at one more example of this type. Consider the following integral:

$$\int 2x\sqrt{(x^2+1)}\,dx$$

Don't be too worried by the fact that we have a root in the integrand. Things are not as complicated as they might appear. In fact, you may well have already realised that the obvious candidate for the substitution is the expression under the root $(x^2 + 1)$, and that the term in front of the radical symbol $(\sqrt{})$ is its derivative, $2x$.

We can therefore make our substitutions and integrate as follows:

$$\int 2x\sqrt{(x^2+1)}\,dx = \int \sqrt{u}\,du$$
$$= \int u^{1/2}\,du$$
$$= \frac{2}{3}(x^2+1)^{3/2} + C$$

Note that we rewrote the root function as a power to enable us to use the power rule for integration. Apart from that, we didn't have to rearrange things at all, since our integral was already in the form:

$$\int f\left(g\left(x\right)\right)g'\left(x\right)dx$$

What we are looking at here is an integrand that is the product of two functions. The first function is a composite function. The second function is the derivative of the composite function's inner function (if the second function happens instead to be a multiple or submultiple of this derivative, we will simply remove the offending multiplier and put it in front of the integral symbol). Once we have our integrand in the required format, we just need to substitute u for *g(x)* and du for *g'(x)* dx so that we are left with:

$$\int f\left(u\right)du$$

We can then carry out integration with respect to *u*, and finish things off by back substituting for u in order to get our answer in terms of *x*.

Integrating the Quotient of Two Functions

When differentiating the quotient of two functions, we can call on the quotient rule. Unfortunately, there is no direct equivalent of this rule when it comes to integrating the quotient of two functions, but we do have several ways of dealing with problems of this nature, some of which involve integration by substitution. Let's start by looking at an example. Consider the following integral:

$$\int \frac{x}{x^2+1}dx$$

Let's assume that we are going to substitute *u* for *x*² + 1. The derivative of *x*² + 1 is 2*x*, so by rearranging things a little bit we can write:

$$\frac{1}{2}\int \frac{2x}{x^2+1}dx$$

We can now make our substitutions (we will need to rearrange things again slightly):

$$\frac{1}{2}\int \frac{1}{u}du$$

Following is the simple rule to find the reciprocal of a variable:

$$\int \frac{1}{x}dx = ln\left(x\right)+C$$

Applying this to our current problem, we get:

$$\frac{1}{2}\int \frac{1}{u}du = \frac{1}{2}ln\left(u\right)+C$$

And back substituting $x^2 + 1$ for u we get:

$$\int \frac{x}{x^2+1}\,dx = \frac{1}{2}\ln\left(x^2+1\right)+C$$

While we're on the subject of reciprocals, consider the following integral:

$$\int \frac{1}{1-2x}\,dx$$

We will substitute u for 1 - 2x. Since the derivative of 1 - 2x is -2, we need to rearrange things slightly to give us the following integral:

$$-\frac{1}{2}\int \frac{1}{u}\,du$$

Now we integrate:

$$-\frac{1}{2}\int \frac{1}{u}\,du = -\frac{1}{2}\ln\left(u\right)+C$$

Back substituting for u gives us:

$$\int \frac{1}{1-2x}\,dx = -\frac{1}{2}\ln\left(1-2x\right)+C$$

The interesting thing about this result is that we can generalise it to enable us to find a solution for any integral in the form:

$$\int \frac{1}{ax+b}\,dx$$

The substitution $u = ax + b$ will leave us with the integral,

$$\frac{1}{a}\int \frac{1}{u}\,du$$

which evaluates to:

$$\frac{1}{a}\ln\left(ax+b\right)+C$$

This means that, given any integration problem in this format, we can write down the solution immediately without having to carry out any intermediate steps. For example, given the following integral:

$$\int \frac{1}{5x+9}\,dx$$

We can immediately write the following result:

$$\int \frac{1}{5x+9}\,dx = \frac{1}{5}\ln\left(5x+9\right)+C$$

Let's look at one more example involving the quotient of two functions. Suppose we want to evaluate the following:

$$\int \frac{4x}{\sqrt{\left(2x^2+1\right)}}\,dx$$

Remember that, as always, we are trying to get our integral into the form:

$$\int f\left(g\left(x\right)\right)\,g'\left(x\right)dx$$

The obvious substitution here is $u = 2x^2 + 1$, and since the derivative of $u = 2x^2 + 1$ is $4x$, we can get our integral into the required format by rearranging it as follows:

$$\int \frac{1}{\sqrt{\left(2x^2+1\right)}}4x\,dx$$

For the sake of clarity, note that $f(u)$, i.e. $f(g(x))$, will give us the reciprocal of $\sqrt{u}$. We can now make our substitutions and evaluate the integral:

$$\int \frac{1}{\sqrt{u}}\,du = \int u^{-1/2}\,du$$
$$= 2u^{-1/2} + C$$

Back substituting for u gives us:

$$\int \frac{4x}{\sqrt{\left(2x^2+1\right)}}\,dx = 2\left(2x^2+1\right)^{1/2} + C$$

Or alternatively:

$$\int \frac{4x}{\sqrt{\left(2x^2+1\right)}}\,dx = 2\sqrt{\left(2x^2+1\right)} + C$$

Using Substitution to Evaluate Definite Integrals

Remember that when we want to evaluate a definite integral, we still need to find the indefinite integral of an expression. Once we have the indefinite integral, however, we will use it to evaluate the value of the resulting function at both the upper and lower limits of integration. The difference between these two values gives us the definite integral, which usually represents the area under the curve of the graph of the function we are integrating between the upper and lower limits of integration.

The limits of integration with respect to the integrand in its original form are values of x. When we undertake u-substitution, we can evaluate the definite integral using the limits of integration without carrying out any back substitution, but we must remember that the limits of integration we use will be expressed in terms of u and not x. Putting this another way, the endpoints we use to evaluate the definite integral will be different. How this works should become clearer once we have looked at a couple of examples.

Consider the following definite integral:

$$\int_1^5 (x+7)^2 \, dx$$

Clearly, the substitution we need to make here will be u = x + 7.

$$du = \left(\frac{du}{dx}\right) dx$$

Since dx always evaluates to one, and the derivative of x + 7 also evaluates to one, we have,

$$\frac{du}{dx} = 1$$

and therefore:

$$du = \left(\frac{du}{dx}\right) dx = dx$$

Our integral can now be rewritten as:

$$\int_{x=1}^{x=5} u^2 \, du$$

We have explicitly written the upper and lower limits of integration to show that they are expressed in terms of x. It is good practice to do this when dealing with problems of this type, and avoids confusion. In order to express the limits of integration in terms of u, we simply apply the substitution to each, as follows:

Upper limit: 7 + x = 7 + 5 = 12;

Lower limit: 7 + x = 7 + 1 = 8.

Now we can rewrite our integral with the limits of integration expressed in terms of u:

$$\int_{u=8}^{u=12} u^2 \, du$$

Integrating will give us:

$$\int_{u=8}^{u=12} u^2 \, du = \left[\frac{1}{3} u^2\right]_8^{12}$$

$$= \frac{1}{3}\left(12^3 - 8^3\right)$$

$$= \frac{1216}{3}$$

Let's try another example. Consider the following integral:

$$\int_0^2 x\cos\left(x^2+1\right)dx$$

Suppose we make the substitution $u = x^2 + 1$. Since the derivative of $x^2 + 1$ is $2x$.

We have:

du = 2x dx

Our integral can therefore be rewritten as follows:

$$\frac{1}{2}\int_{u=1}^{u=5}\cos(u)\,du$$

Integrating will give us:

$$\frac{1}{2}\int_{u=1}^{u=5}\cos(u)\,du = \frac{1}{2}\Big[\cos(u)\Big]_1^5$$

$$= \frac{1}{2}\big(\cos(5)-\cos(1)\big)$$

Integration by Parts

In calculus, and more generally in mathematical analysis, integration by parts or partial integration is a process that finds the integral of a product of functions in terms of the integral of the product of their derivative and antiderivative. It is frequently used to transform the antiderivative of a product of functions into an antiderivative for which a solution can be more easily found. The rule can be thought of as an integral version of the product rule of differentiation.

If $u = u(x)$ and $du = u'(x)\,dx$, while $v = v(x)$ and $dv = v'(x)\,dx$, then integration by parts states that,

$$\int_a^b u(x)v'(x)dx = \Big[u(x)v(x)\Big]_a^b - \int_a^b u'(x)v(x)dx$$

$$= u(b)v(b) - u(a)v(a) - \int_a^b u'(x)v(x)dx$$

or more compactly:

$$\int u\,dv = uv - \int v\,du.$$

Mathematician Brook Taylor discovered integration by parts, first publishing the idea in 1715. More general formulations of integration by parts exist for the Riemann–Stieltjes and Lebesgue–Stieltjes integrals. The discrete analogue for sequences is called summation by parts.

Theorem: Product of Two Functions

The theorem can be derived as follows. For two continuously differentiable functions $u(x)$ and $v(x)$, the product rule states:

$$\big(u(x)v(x)\big)' = v(x)u'(x) + u(x)v'(x).$$

Integrating both sides with respect to x,

$$\int \big(u(x)v(x)\big)' dx = \int u'(x)v(x)dx + \int u(x)v'(x)dx,$$

and noting that an indefinite integral is an antiderivative gives,

$$u(x)v(x) = \int u'(x)v(x)dx + \int u(x)v'(x)dx,$$

where we neglect writing the constant of integration. This yields the formula for integration by parts:

$$\int u(x)v'(x)dx = u(x)v(x) - \int u'(x)v(x)dx,$$

or in terms of the differentials $du = u'(x)dx$, $dv = v'(x)dx$,

$$\int u(x)dv = u(x)v(x) - \int v(x)du.$$

This is to be understood as an equality of functions with an unspecified constant added to each side. Taking the difference of each side between two values $x = a$ and $x = b$ and applying the fundamental theorem of calculus gives the definite integral version:

$$\int_a^b u(x)v'(x)dx = u(b)v(b) - u(a)v(a) - \int_a^b u'(x)v(x)dx.$$

The original integral $\int uv' \, dx$ contains the derivative v'; to apply the theorem, one must find v, the antiderivative of v', then evaluate the resulting integral $\int vu' \, dx$.

Validity for Less Smooth Functions

It is not necessary for u and v to be continuously differentiable. Integration by parts works if u is absolutely continuous and the function designated v' is Lebesgue integrable (but not necessarily continuous). (If v' has a point of discontinuity then its antiderivative v may not have a derivative at that point).

If the interval of integration is not compact, then it is not necessary for u to be absolutely continuous in the whole interval or for v' to be Lebesgue integrable in the interval, as a couple of examples (in which u and v are continuous and continuously differentiable) will show. For instance,

$$u(x) = \exp(x)/x^2, v'(x) = \exp(-x)$$

u is not absolutely continuous on the interval $[1, \infty)$, but nevertheless,

$$\int_1^\infty u(x)v'(x)dx = \left[u(x)v(x)\right]_1^\infty - \int_1^\infty u'(x)v(x)dx$$

so long as $\left[u(x)v(x)\right]_1^\infty$ is taken to mean the limit of $u(L)v(L) - u(1)v(1)$ as $L \to \infty$ and so long as the two terms on the right-hand side are finite. This is only true if we choose $v(x) = -\exp(-x)$.

Similarly, if,

$$u(x) = \exp(-x), v'(x) = x^{-1}\sin(x)$$

v' is not Lebesgue integrable on the interval $[1, \infty)$, but nevertheless,

$$\int_1^\infty u(x)v'(x)dx = \left[u(x)v(x)\right]_1^\infty - \int_1^\infty u'(x)v(x)dx$$

with the same interpretation.

One can also easily come up with similar examples in which u and v are *not* continuously differentiable. Further, if $f(x)$ is a function of bounded variation on the segment $[a,b]$, and $\varphi(x)$ is differentiable on $[a,b]$, then,

$$\int_a^b f(x)\varphi'(x)dx = -\int_{-\infty}^\infty \tilde{\varphi}(x)d(\tilde{\chi}_{[a,b]}(x)\tilde{f}(x)),$$

where $d(\chi_{[a,b]}(x)f(x))$ denotes the signed measure corresponding to the function of bounded variation $\chi_{[a,b]}(x)f(x)$, and functions $\tilde{f}, \tilde{\varphi}$ are extensions of f, φ to $\mathbb{R}$, which are respectively of bounded variation and differentiable.

Product of many Functions

Integrating the product rule for three multiplied functions, $u(x)$, $v(x)$, $w(x)$, gives a similar result:

$$\int_a^b uv\,dw = \left[uvw\right]_a^b - \int_a^b uw\,dv - \int_a^b vw\,du.$$

In general, for n factors,

$$\left(\prod_{i=1}^n u_i(x)\right)' = \sum_{j=1}^n \prod_{i \neq j}^n u_i(x)u_j'(x),$$

which leads to,

$$\left[\prod_{i=1}^{n} u_i(x)\right]_a^b = \sum_{j=1}^{n} \int_a^b \prod_{i \neq j}^{n} u_i(x) u_j'(x),$$

where the product is of all functions except for the one differentiated in the same term.

Visualization

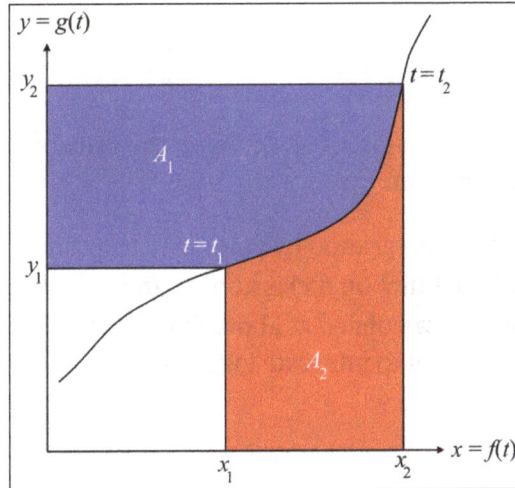

The pictured curve is parametrized by the variable t.

Consider a parametric curve by $(x, y) = (f(t), g(t))$. Assuming that the curve is locally one-to-one and integrable, we can define:

$$x(y) = f(g^{-1}(y))$$
$$y(x) = g(f^{-1}(x))$$

The area of the blue region is:

$$A_1 = \int_{y_1}^{y_2} x(y)\,dy$$

Similarly, the area of the red region is:

$$A_2 = \int_{x_1}^{x_2} y(x)\,dx$$

The total area $A_1 + A_2$ is equal to the area of the bigger rectangle, $x_2 y_2$, minus the area of the smaller one, $x_1 y_1$:

$$\overbrace{\int_{y_1}^{y_2} x(y)\,dy}^{A_1} + \overbrace{\int_{x_1}^{x_2} y(x)\,dx}^{A_2} = x \cdot y(x)\Big|_{x_1}^{x_2} = y \cdot x(y)\Big|_{y_1}^{y_2}.$$

Or, in terms of t:

$$\int_{t_1}^{t_2} x(t)dy(t) + \int_{t_1}^{t_2} y(t)dx(t) = x(t)y(t)\Big|_{t_1}^{t_2}$$

Or, in terms of indefinite integrals, this can be written as:

$$\int x\,dy + \int y\,dx = xy$$

Rearranging:

$$\int x\,dy = xy - \int y\,dx$$

Thus integration by parts may be thought of as deriving the area of the blue region from the area of rectangles and that of the red region.

This visualization also explains why integration by parts may help find the integral of an inverse function $f^{-1}(x)$ when the integral of the function $f(x)$ is known. Indeed, the functions $x(y)$ and $y(x)$ are inverses, and the integral $\int x\,dy$ may be calculated as above from knowing the integral $\int y\,dx$. In particular, this explains use of integration by parts to integrate logarithm and inverse trigonometric functions.

Applications

Finding Antiderivatives

Integration by parts is a heuristic rather than a purely mechanical process for solving integrals; given a single function to integrate, the typical strategy is to carefully separate this single function into a product of two functions $u(x)v(x)$ such that the residual integral from the integration by parts formula is easier to evaluate than the single function. The following form is useful in illustrating the best strategy to take:

$$\int uv\,dx = u \int v\,dx - \int \left(u' \int v\,dx\right) dx.$$

On the right-hand side, u is differentiated and v is integrated; consequently it is useful to choose u as a function that simplifies when differentiated, or to choose v as a function that simplifies when integrated. As a simple example, consider:

$$\int \frac{\ln(x)}{x^2}\,dx.$$

Since the derivative of $\ln(x)$ is $\frac{1}{x}$, one makes $(\ln(x))$ part u; since the antiderivative of $\frac{1}{x^2}$ is $-\frac{1}{x}$, one makes $\frac{1}{x^2}\,dx$ part dv. The formula now yields:

$$\int \frac{\ln(x)}{x^2}\,dx = -\frac{\ln(x)}{x} - \int \left(\frac{1}{x}\right)\left(-\frac{1}{x}\right) dx.$$

The antiderivative of $-\frac{1}{x^2}$ can be found with the power rule and is $\frac{1}{x}$.

Alternatively, one may choose u and v such that the product u' ($\int v\, dx$) simplifies due to cancellation. For example, suppose one wish to integrate:

$$\int \sec^2(x) \cdot \ln\left(\left|\sin(x)\right|\right) dx.$$

If we choose $u(x) = \ln(|\sin(x)|)$ and $v(x) = \sec^2 x$, then u differentiates to $1/\tan x$ using the chain rule and v integrates to $\tan x$; so the formula gives:

$$\int \sec^2(x) \cdot \ln(|\sin(x)|)\, dx = \tan(x) \cdot \ln(|\sin(x)|) - \int \tan(x) \cdot \frac{1}{\tan(x)} dx \ .$$

The integrand simplifies to 1, so the antiderivative is x. Finding a simplifying combination frequently involves experimentation.

In some applications, it may not be necessary to ensure that the integral produced by integration by parts has a simple form; for example, in numerical analysis, it may suffice that it has small magnitude and so contributes only a small error term.

Polynomials and Trigonometric Functions

In order to calculate,

$$I = \int x \cos(x)\, dx\ ,$$

let,

$$u = x \Rightarrow du = dx$$

$$dv = \cos(x)\, dx \Rightarrow v = \int \cos(x)\, dx = \sin(x)$$

then,

$$\int x \cos(x)\, dx = \int u\, dv$$

$$= u \cdot v - \int v\, du$$

$$= x \sin(x) - \int \sin(x)\, dx$$

$$= x \sin(x) + \cos(x) + C,$$

where C is a constant of integration.

For higher powers of x in the form:

$$\int x^n e^x\, dx, \int x^n \sin(x)\, dx, \int x^n \cos(x)\, dx,$$

Repeatedly using integration by parts can evaluate integrals such as these; each application of the theorem lowers the power of x by one.

Exponentials and Trigonometric Functions

An example commonly used to examine the workings of integration by parts is:

$$I = \int e^x \cos(x)\, dx.$$

Here, integration by parts is performed twice.

First let,

$$u = \cos(x) \Rightarrow du = -\sin(x)\, dx$$
$$dv = e^x\, dx \Rightarrow v = \int e^x\, dx = e^x$$

then,

$$\int e^x \cos(x)\, dx = e^x \cos(x) + \int e^x \sin(x)\, dx.$$

Now, to evaluate the remaining integral, we use integration by parts again, with:

$$u = \sin(x) \Rightarrow du = \cos(x)\, dx$$
$$dv = e^x\, dx \Rightarrow v = \int e^x\, dx = e^x.$$

Then,

$$\int e^x \sin(x)\, dx = e^x \sin(x) - \int e^x \cos(x)\, dx.$$

Putting these together:

$$\int e^x \cos(x)\, dx = e^x \cos(x) + e^x \sin(x) - \int e^x \cos(x)\, dx.$$

The same integral shows up on both sides of this equation. The integral can simply be added to both sides to get,

$$2 \int e^x \cos(x)\, dx = e^x \big(\sin(x) + \cos(x) \big) + C$$

which rearranges to,

$$\int e^x \cos(x)\, dx = \frac{e^x \big(\sin(x) + \cos(x) \big)}{2} + C'$$

where again C (and $C' = C/2$) is a constant of integration. A similar method is used to find the integral of secant cubed.

Functions Multiplied by Unity

Two other well-known examples are when integration by parts is applied to a function expressed as a product of 1 and itself. This works if the derivative of the function is known, and the integral of this derivative times x is also known.

The first example is $\int \ln(x)\, dx$. We write this as:

$$I = \int \ln(x) \cdot 1\, dx.$$

Let,

$$u = \ln(x) \Rightarrow du = \frac{dx}{x}$$

$$dv = dx \Rightarrow v = x$$

then,

$$\int \ln(x)\, dx = x\ln(x) - \int \frac{x}{x}\, dx$$
$$= x\ln(x) - \int 1\, dx$$
$$= x\ln(x) - x + C$$

where C is the constant of integration.

The second example is the inverse tangent function $\arctan(x)$:

$$I = \int \arctan(x)\, dx.$$

Rewrite this as,

$$\int \arctan(x) \cdot 1\, dx$$

Now let,

$$u = \arctan(x) \Rightarrow du = \frac{dx}{1+x^2}$$

$$dv = dx \Rightarrow v = x$$

then,

$$\int \arctan(x)\, dx = x\arctan(x) - \int \frac{x}{1+x^2}\, dx$$
$$= x\arctan(x) - \frac{\ln(1+x^2)}{2} + C$$

using a combination of the inverse chain rule method and the natural logarithm integral condition.

LIATE Rule

A rule of thumb has been proposed, consisting of choosing as u the function that comes first in the following list:

- L – logarithmic functions: $\ln(x)$, $\log_b(x)$ etc.
- I – inverse trigonometric functions: $\arctan(x)$, $\text{arcsec}(x)$, etc.
- A – algebraic functions: x^2, $3x^{50}$, etc.
- T – trigonometric functions: $\sin(x)$, $\tan(x)$, etc.
- E – exponential functions: e^x, 19^x, etc.

The function which is to be *dv* is whichever comes last in the list: functions lower on the list have easier antiderivatives than the functions above them. The rule is sometimes written as "DETAIL" where *D* stands for *dv*.

To demonstrate the LIATE rule, consider the integral:

$$\int x \cdot \cos(x) dx.$$

Following the LIATE rule, $u = x$, and $dv = \cos(x)\, dx$, hence $du = dx$, and $v = \sin(x)$, which makes the integral become,

$$x \cdot \sin(x) - \int 1 \sin(x) dx,$$

which equals:

$$x \cdot \sin(x) + \cos(x) + C.$$

In general, one tries to choose *u* and *dv* such that *du* is simpler than *u* and *dv* is easy to integrate. If instead $\cos(x)$ was chosen as *u*, and xdx as *dv*, we would have the integral,

$$\frac{x^2}{2}\cos(x) + \int \frac{x^2}{2}\sin(x) dx,$$

which, after recursive application of the integration by parts formula, would clearly result in an infinite recursion and lead nowhere.

Although a useful rule of thumb, there are exceptions to the LIATE rule. A common alternative is to consider the rules in the "ILATE" order instead. Also, in some cases, polynomial terms need to be split in non-trivial ways. For example, to integrate,

$$\int x^3 e^{x^2}\, dx,$$

one would set,

$$u = x^2, \quad dv = x \cdot e^{x^2}\, dx,$$

so that,

$$du = 2xdx, \quad v = \frac{e^{x^2}}{2}.$$

Then,

$$\int x^3 e^{x^2}\, dx = \int (x^2)(xe^{x^2}) dx = \int u\, dv = uv - \int v\, du = \frac{x^2 e^{x^2}}{2} - \int xe^{x^2}\, dx.$$

Finally, this results in:

$$\int x^3 e^{x^2}\, dx = \frac{e^{x^2}(x^2 - 1)}{2} + C.$$

Integration by parts is often used as a tool to prove theorems in mathematical analysis.

Gamma Function Identity

The gamma function is an example of a special function, defined as an improper integral for $z > 0$. Integration by parts illustrates it to be an extension of the factorial:

$$\Gamma(z) = \int_0^\infty e^{-x} x^{z-1} dx$$

$$= -\int_0^\infty x^{z-1} d\left(e^{-x}\right)$$

$$= -\left[e^{-x} x^{z-1}\right]_0^\infty + \int_0^\infty e^{-x} d\left(x^{z-1}\right)$$

$$= 0 + \int_0^\infty (z-1) x^{z-2} e^{-x} dx$$

$$= (z-1)\Gamma(z-1).$$

Since,

$$\Gamma(1) = \int_0^\infty e^{-x} dx = 1,$$

for integer z, applying this formula repeatedly gives the factorial:

$$\Gamma(z+1) = z!$$

Use in Harmonic Analysis

Integration by parts is often used in harmonic analysis, particularly Fourier analysis, to show that quickly oscillating integrals with sufficiently smooth integrands decay quickly. The most common example of this is its use in showing that the decay of function's Fourier transform depends on the smoothness of that function.

Fourier Transform of Derivative

If f is a k-times continuously differentiable function and all derivatives up to the kth one decay to zero at infinity, then its Fourier transform satisfies,

$$(\mathcal{F}f^{(k)})(\xi) = (2\pi i\xi)^k \mathcal{F}f(\xi),$$

where $f^{(k)}$ is the kth derivative of f. (The exact constant on the right depends on the convention of the Fourier transform used.) This is proved by noting that,

$$\frac{d}{dy} e^{-2\pi i y\xi} = -2\pi i\xi e^{-2\pi i y\xi},$$

so using integration by parts on the Fourier transform of the derivative we get,

$$(Ff')(\xi) = \int_{-\infty}^{\infty} e^{-2\pi i y \xi} f'(y) dy$$

$$= [e^{-2\pi i y \xi} f(y)]_{-\infty}^{\infty} - \int_{-\infty}^{\infty} (-2\pi i \xi e^{-2\pi i y \xi}) f(y) dy$$

$$= 2\pi i \xi \int_{-\infty}^{\infty} e^{-2\pi i y \xi} f(y) dy$$

$$= 2\pi i \xi Ff(\xi).$$

Applying this inductively gives the result for general k. A similar method can be used to find the Laplace transform of a derivative of a function.

Decay of Fourier Transform

The above result tells us about the decay of the Fourier transform, since it follows that if f and $f^{(k)}$ are integrable then:

$$|Ff(\xi)| \le \frac{I(f)}{1+|2\pi\xi|^k}, \text{ where } I(f) = \int_{-\infty}^{\infty} \left(|f(y)| + |f^{(k)}(y)| \right) dy.$$

In other words, if f satisfies these conditions then its Fourier transform decays at infinity at least as quickly as $1/|\xi|^k$. In particular, if $k \ge 2$ then the Fourier transform is integrable.

The proof uses the fact, which is immediate from the definition of the Fourier transform, that:

$$|Ff(\xi)| \le \int_{-\infty}^{\infty} |f(y)| dy.$$

Using the same idea on the equality stated at the start of this subsection gives:

$$|(2\pi i \xi)^k Ff(\xi)| \le \int_{-\infty}^{\infty} |f^{(k)}(y)| dy.$$

Summing these two inequalities and then dividing by $1 + |2\pi \xi^k|$ gives the stated inequality.

Use in Operator Theory

One use of integration by parts in operator theory is that it shows that the $-\Delta$ (where Δ is the Laplace operator) is a positive operator on L^2. If f is smooth and compactly supported then, using integration by parts we have,

$$\langle -\Delta f, f \rangle_{L^2} = -\int_{-\infty}^{\infty} f''(x) \overline{f(x)} dx$$

$$= -\left[f'(x) \overline{f(x)} \right]_{-\infty}^{\infty} + \int_{-\infty}^{\infty} f'(x) \overline{f'(x)} dx$$

$$= \int_{-\infty}^{\infty} |f'(x)|^2 dx \ge 0.$$

Other Applications

- Determining boundary conditions in Sturm–Liouville theory.

- Deriving the Euler–Lagrange equation in the calculus of variations.

Repeated Integration by Parts

Considering a second derivative of v in the integral on the LHS of the formula for partial integration suggests a repeated application to the integral on the RHS:

$$\int uv'' \, dx = uv' - \int u'v' \, dx = uv' - \left(u'v - \int u''v \, dx \right).$$

Extending this concept of repeated partial integration to derivatives of degree n leads to:

$$\int u^{(0)} v^{(n)} \, dx = u^{(0)} v^{(n-1)} - u^{(1)} v^{(n-2)} + u^{(2)} v^{(n-3)} - \cdots + (-1)^{n-1} u^{(n-1)} v^{(0)} + (-1)^n \int u^{(n)} v^{(0)} \, dx.$$

$$= \sum_{k=0}^{n-1} (-1)^k u^{(k)} v^{(n-1-k)} + (-1)^n \int u^{(n)} v^{(0)} \, dx.$$

This concept may be useful when the successive integrals of $v^{(n)}$ are readily available (e.g., plain exponentials or sine and cosine, as in Laplace or Fourier transforms), and when the nth derivative of u vanishes (e.g., as a polynomial function with degree $(n-1)$). The latter condition stops the repeating of partial integration, because the RHS-integral vanishes.

In the course of the above repetition of partial integrations the integrals:

$$\int u^{(0)} v^{(n)} \, dx \quad \text{and} \quad \int u^{(\ell)} v^{(n-\ell)} \, dx \quad \text{and}$$

$$\int u^{(m)} v^{(n-m)} \, dx \quad \text{for } 1 \le m, \ell \le n$$

get related. This may be interpreted as arbitrarily "shifting" derivatives between v and u within the integrand, and proves useful, too.

Tabular Integration by Parts

The essential process of the above formula can be summarized in a table; the resulting method is called "tabular integration" and was featured in the film Stand and Deliver.

For example, consider the integral:

$$\int x^3 \cos x \, dx \quad \text{and take } u^{(0)} = x^3, \quad v^{(n)} = \cos x.$$

Begin to list in column A the function $u^{(0)} = x^3$ and its subsequent derivatives $u^{(i)}$ until zero is reached. Then list in column B the function $v^{(n)} = \cos x$ and its subsequent integrals $v^{(n-i)}$ until the size of column B is the same as that of column A. The result is as follows:

i	Sign	A: derivatives $u^{(i)}$	B: integrals $v^{(n-i)}$
0	+	x^3	$\cos x$

1	−	$3x^2$	$\sin x$
2	+	$6x$	$-\cos x$
3	−	6	$-\sin x$
4	+	0	$\cos x$

The product of the entries in row i of columns A and B together with the respective sign give the relevant integrals in step i in the course of repeated integration by parts. Step $i = 0$ yields the original integral. For the complete result in step $i > 0$ the ith integral must be added to all the previous products ($0 \leq j < i$) of the jth entry of column A and the $(j + 1)$st entry of column B (i.e., multiply the 1st entry of column A with the 2nd entry of column B, the 2nd entry of column A with the 3rd entry of column B, etc.) with the given jth sign. This process comes to a natural halt, when the product, which yields the integral, is zero ($i = 4$ in the example). The complete result is the following (with the alternating signs in each term):

$$\underbrace{(+1)(x^3)(\sin x)}_{j=0} + \underbrace{(-1)(3x^2)(-\cos x)}_{j=1} + \underbrace{(+1)(6x)(-\sin x)}_{j=2} + \underbrace{(-1)(6)(\cos x)}_{j=3} + \underbrace{\int(+1)(0)(\cos x)dx}_{i=4: \to C}$$

This yields,

$$\underbrace{\int x^3 \cos x\,dx}_{\text{step 0}} = x^3 \sin x + 3x^2 \cos x - 6x\sin x - 6\cos x + C.$$

The repeated partial integration also turns out useful, when in the course of respectively differentiating and integrating the functions $u^{(i)}$ and $v^{(n-i)}$ their product results in a multiple of the original integrand. In this case the repetition may also be terminated with this index i. This can happen, expectably, with exponentials and trigonometric functions. As an example consider, $\int e^x \cos x\,dx$.

i	Sign	A: derivatives $u^{(i)}$	B: integrals $v^{(n-i)}$
0	+	e^x	$\cos x$
1	−	e^x	$\sin x$
2	+	e^x	$-\cos x$

In this case the product of the terms in columns A and B with the appropriate sign for index $i = 2$ yields the negative of the original integrand (compare rows $i = 0$ and $i = 2$).

$$\underbrace{\int e^x \cos x\,dx}_{\text{step 0}} = \underbrace{(+1)(e^x)(\sin x)}_{j=0} + \underbrace{(-1)(e^x)(-\cos x)}_{j=1} + \underbrace{\int(+1)(e^x)(-\cos x)dx}_{i=2}.$$

Observing that the integral on the RHS can have its own constant of integration C', and bringing the abstract integral to the other side, gives,

$$2\int e^x \cos x\,dx = e^x \sin x + e^x \cos x + C',$$

and finally,

$$\int e^x \cos x dx = \frac{1}{2}(e^x(\sin x + \cos x)) + C,$$

where $C = C'/2$.

Higher Dimensions

Integration by parts can be extended to functions of several variables by applying a version of the fundamental theorem of calculus to an appropriate product rule. There are several such pairings possible in multivariate calculus, involving a scalar-valued function u and vector-valued function (vector field) V.

The product rule for divergence states:

$$\nabla \cdot (uV) = u\nabla \cdot V + \nabla u \cdot V.$$

Suppose Ω is an open bounded subset of $\mathbb{R}^n$ with a piecewise smooth boundary $\Gamma = \partial\Omega$ Integrating over Ω with respect to the standard volume form $d\Omega$, and applying the divergence theorem, gives,

$$\int_\Gamma uV \cdot \hat{n} d\Gamma = \int_\Omega \nabla \cdot (uV) \, d\Omega = \int_\Omega u\nabla \cdot V d\Omega + \int_\Omega \nabla u \cdot V d\Omega,$$

where $\hat{n}$ is the outward unit normal vector to the boundary, integrated with respect to its standard Riemannian volume form $d\Gamma$. Rearranging gives,

$$\int_\Omega u\nabla \cdot V d\Omega = \int_\Gamma uV \cdot \hat{n} d\Gamma - \int_\Omega \nabla u \cdot V d\Omega,$$

or in other words,

$$\int_\Omega u \, \text{div}\,(V) \, d\Omega = \int_\Gamma uV \cdot \hat{n} d\Gamma - \int_\Omega \text{grad}(u) \cdot V d\Omega.$$

The regularity requirements of the theorem can be relaxed. For instance, the boundary $\Gamma = \partial\Omega$ need only be Lipschitz continuous, and the functions u, v need only lie in the Sobolev space $H^1(\Omega)$.

First Green's Identity

Consider the continuously differentiable vector fields $U = u_1 e_1 + \cdots + u_n e_n$ and $ve_1, \ldots, ve_n$, where u_i is the i-th standard basis vector for $i = 1, \ldots, n.$. Now apply the above integration by parts to each u_i times the vector field ve_i:

$$\int_\Omega u_i \frac{\partial v}{\partial x_i} d\Omega = \int_\Gamma u_i \, ve_i \cdot \hat{n} d\Gamma - \int_\Omega \frac{\partial u_i}{\partial x_i} v d\Omega.$$

Summing over i gives a new integration by parts formula:

$$\int_\Omega U \cdot \nabla v d\Omega = \int_\Gamma U \cdot \hat{n} d\Gamma - \int_\Omega v \nabla \cdot U d\Omega.$$

The case $U = \nabla u,$, where $u \in C^2(\bar{\Omega}),$, is known as the first Green's identity:

$$\int_\Omega \nabla u \cdot \nabla v d\Omega = \int_\Gamma v \nabla u \cdot \hat{n} d\Gamma - \int_\Omega v \nabla^2 u d\Omega.$$

Integration by Trogonometric Substitution

The following integration problems use the method of trigonometric (trig) substitution. It is a method for finding antiderivatives of functions which contain square roots of quadratic expressions or rational powers of the form $\frac{n}{2}$ (where n is an integer) of quadratic expressions. Examples of such expressions are:

$$\sqrt{4-x^2} \ and \ \left(x^2+1\right)^{3/2}$$

The method of trig substitution may be called upon when other more common and easier-to-use methods of integration have failed. Trig substitution assumes that you are familiar with standard trigonometric identies, the use of differential notation, integration using u-substitution, and the integration of trigonometric functions:

$$x = f(\theta),$$
$$dx = f'(\theta) \, d\theta$$

For example, if:

$$x = \sec \theta,$$

then,

$$dx = \sec \theta \tan \theta \, d\theta$$

The goal of trig substitution will be to replace square roots of quadratic expressions or rational powers of the form $\frac{n}{2}$ (where n is an integer) of quadratic expressions, which may be impossible to integrate using other methods of integration, with integer powers of trig functions, which are more easily integrated. For example, if we start with the expression,

$$\sqrt{4-x^2}$$

and let,

$$x = 2\sin \theta,$$

then,

$$\sqrt{4-x^2} = \sqrt{4-\left(2\sin \theta\right)^2}$$

ng the given right triangle and the Pythagorean Theorem, we can determine any trig value of θ.

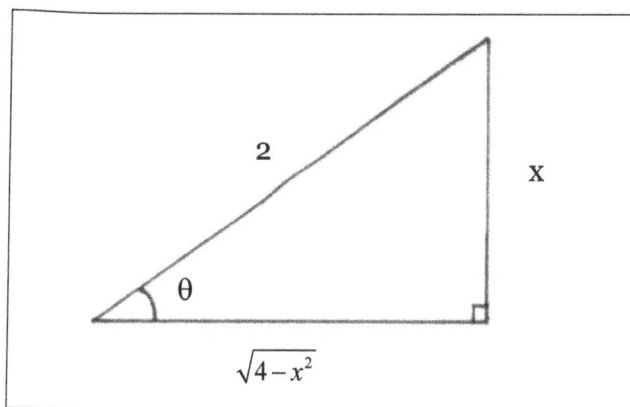

$(adjacent)^2 + (opposite)^2 = (hypotenuse)^2 \rightarrow$

$(adjacent)^2 + (x)^2 = (2)^2 \rightarrow adjacent = \sqrt{4-x^2} \rightarrow$

$\cos \dfrac{adjacent}{hypotenuse} \qquad \dfrac{\sqrt{} \quad x}{}$

$2\theta + 2\sin\theta\cos\theta + C = 2\arcsin\left(\dfrac{x}{2}\right) + 2 \cdot \dfrac{x}{2} \cdot \dfrac{\sqrt{4-x^2}}{2}$

$= 2\arcsin\left(\dfrac{x}{2}\right) + \dfrac{1}{2}x \cdot \sqrt{4-x^2} + C$

sing the method of trig substitution, we will always use one of the following three well-
rig identities:

$- \sin^2\theta = \cos^2\theta$

$+ \tan^2\theta = \sec^2\theta \ and$

$ec\,2\theta - 1 = \tan 2\theta$

xpression,

$\overline{a^2 - x^2}$

uation $1 - \sin^2\theta = \cos^2\theta$ and let,

$= a\sin\theta$

at $-\dfrac{\pi}{2} \le \theta \le \dfrac{\pi}{2}$ so that $\cos\theta \ge 0$. This allows for both positive and negative values of x.

$$= \sqrt{4 - 4\sin^2\theta}$$
$$= \sqrt{4\left(1 - \sin^2\theta\right)}$$
$$= \sqrt{4} \cdot \sqrt{1 - \sin^2\theta}$$

(Recall that cos²θ + sin²θ = 1 so that 1 − sin² θ = cos² θ.)

$$= 2 \cdot \sqrt{\cos^2\theta}$$
$$= 2 \cdot |\cos\theta|$$

(Assume that $-\dfrac{\pi}{2} \le \theta \le \dfrac{\pi}{2}$ so that cos θ ≥ 0.)

$$= 2\cos\theta$$

and

$$dx = 2\cos\theta \, d\theta$$

thus,

$$\int \sqrt{4 - x^2} \, dx$$

could be rewritten as:

$$\int \sqrt{4 - x^2} \, dx = \int 2\cos\theta \cdot 2\cos\theta d\theta = 4\int \cos^2\theta \, d\theta$$

Recall that $\cos 2\theta = 2\cos^2\theta - 1$ so that $\cos^2\theta = \dfrac{1}{2}(1 + \cos 2\theta)$.

$$= 4\int \frac{1}{2}(1 + \cos 2\theta) d\theta$$
$$= 2\int (1 + \cos 2\theta) d\theta$$
$$= 2\left(\theta + \frac{1}{2}\sin 2\theta\right) + C$$
$$= 2\theta + \sin 2\theta + C$$

Recall that sin 2θ = 2 sin θ cos θ.

$$= 2\theta + 2\sin\theta\cos\theta + C$$

We need to write our final answer in terms of x. Since x = 2 sin θ, it

$$\sin\theta = \frac{x}{2} = \frac{opposite}{hypotenuse}$$

and

$$\theta = \arcsin\left(\frac{x}{2}\right)$$

$$= \sqrt{4 - 4\sin^2 \theta}$$

$$= \sqrt{4(1 - \sin^2 \theta)}$$

$$= \sqrt{4} \cdot \sqrt{1 - \sin^2 \theta}$$

(Recall that cos²θ + sin²θ = 1 so that 1 − sin² θ = cos² θ.)

$$= 2 \cdot \sqrt{\cos^2 \theta}$$

$$= 2 \cdot |\cos \theta|$$

(Assume that $-\dfrac{\pi}{2} \le \theta \le \dfrac{\pi}{2}$ so that cos θ ≥ 0.)

$$= 2 \cos \theta$$

and

$$dx = 2\cos \theta \, d\theta$$

thus,

$$\int \sqrt{4 - x^2} \, dx$$

could be rewritten as:

$$\int \sqrt{4 - x^2} \, dx = \int 2\cos \theta \cdot 2\cos \theta \, d\theta = 4 \int \cos^2 \theta \, d\theta$$

Recall that $\cos 2\theta = 2\cos^2 \theta - 1$ so that $\cos^2 \theta = \dfrac{1}{2}(1 + \cos 2\theta)$.

$$= 4 \int \frac{1}{2}(1 + \cos 2\theta) \, d\theta$$

$$= 2 \int (1 + \cos 2\theta) \, d\theta$$

$$= 2\left(\theta + \frac{1}{2}\sin 2\theta \right) + C$$

$$= 2\theta + \sin 2\theta + C$$

Recall that sin 2θ = 2 sin θ cos θ.

$$= 2\theta + 2\sin \theta \cos \theta + C$$

We need to write our final answer in terms of x. Since x = 2 sin θ, it follows that,

$$\sin \theta = \frac{x}{2} = \frac{opposite}{hypotenuse}$$

and

$$\theta = \arcsin\left(\frac{x}{2} \right)$$

Using the given right triangle and the Pythagorean Theorem, we can determine any trig value of θ.

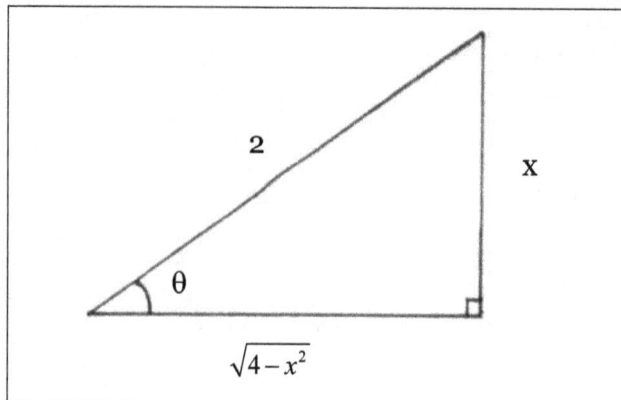

Since,

$$(adjacent)^2 + (opposite)^2 = (hypotenuse)^2 \rightarrow$$

$$(adjacent)^2 + (x)^2 = (2)^2 \rightarrow \quad adjacent = \sqrt{4-x^2} \rightarrow$$

$$\cos \frac{adjacent}{hypotenuse} \quad \frac{\sqrt{} \quad x}{}$$

Then,

$$2\theta + 2\sin\theta\cos\theta + C = 2\arcsin\left(\frac{x}{2}\right) + 2\cdot\frac{x}{2}\cdot\frac{\sqrt{4-x^2}}{2}$$

$$= 2\arcsin\left(\frac{x}{2}\right) + \frac{1}{2}x\cdot\sqrt{4-x^2} + C$$

When using the method of trig substitution, we will always use one of the following three well-known trig identities:

$$1 - \sin^2\theta = \cos^2\theta$$
$$1 + \tan^2\theta = \sec^2\theta \; and$$
$$\sec 2\theta - 1 = \tan 2\theta$$

For the expression,

$$\sqrt{a^2 - x^2}$$

we use equation $1 - \sin^2\theta = \cos^2\theta$ and let,

$$x = a\sin\theta$$

Assume that $-\frac{\pi}{2} \le \theta \le \frac{\pi}{2}$ so that $\cos\theta \ge 0$. This allows for both positive and negative values of x.

Then,

$$\sqrt{a^2 - x^2} = \sqrt{a^2 - a^2 \sin^2 \theta}$$
$$= \sqrt{a^2 \left(1 - \sin^2 \theta\right)}$$
$$= \sqrt{a^2 \cos^2 \theta}$$
$$= \sqrt{a^2} \sqrt{\cos^2 \theta}$$
$$= a \left|\cos \theta\right|$$
$$= a \cos \theta$$

and

$$dx = a \cos \theta \, d\theta$$

For the expression,

$$\sqrt{a^2 + x^2}$$

we use equation $1 + \tan^2 \theta = \sec^2 \theta$ and let,

$$x = a \tan \theta$$

Assume that $-\dfrac{\pi}{2} \le \theta \le \dfrac{\pi}{2}$ so that $\cos \theta > 0$ and $\sec \theta > 0$. This allows for both positive and negative value of x.

Then,

$$\sqrt{a^2 + x^2} = \sqrt{a^2 + a^2 \tan^2 \theta}$$
$$= \sqrt{a^2 \left(1 + \tan^2 \theta\right)}$$

$$= \sqrt{a^2} \sqrt{\sec^2 \theta}$$
$$= a \sqrt{\sec^2 \theta}$$
$$= a \left|\sec \theta\right|$$
$$= a \sec \theta$$

and,

$$dx = a \sec^2 \theta \, d\theta$$

For the expression,

$$\sqrt{x^2 - a^2}$$

we use equation $\sec 2\theta - 1 = \tan 2\theta$ and let,

$$x = a \sec \theta$$

Assume that $0 \le \theta < \dfrac{\pi}{2}$, so that $\tan \theta \ge 0$. This allows for only positive values of x. If the integral includes negative values of x, you must use $\dfrac{\pi}{2} < \theta \le \pi$ with $\sqrt{\tan^2 \theta} = -\tan \theta$.

Then,

$$
\begin{aligned}
\sqrt{x^2 + a^2} &= \sqrt{a^2 \sec^2 \theta - a^2} \\
&= \sqrt{a^2 \left(\sec^2 \theta - 1\right)} \\
&= \sqrt{a^2}\sqrt{\sec^2 \theta - 1} \\
&= a\sqrt{\tan^2 \theta} \\
&= a|\tan \theta| \\
&= a\,\tan \theta
\end{aligned}
$$

and,

$$dx = a \sec \theta \tan \theta \, d\theta$$

Recall the following well-known, basic indefinite trigonometric integral formulas:

1. $\displaystyle\int \cos x \, dx = \sin x + C$

2. $\displaystyle\int \sin x \, dx = -\cos x + C$

3. $\displaystyle\int \sec^2 x \, dx = \tan x + C$

4. $\displaystyle\int \csc^2 x \, dx = -\cot x + C$

5. $\displaystyle\int \sec x \tan x \, dx = \sec x + C$

6. $\displaystyle\int \csc x \cot x \, dx = -\csc x + C$

7. $\displaystyle\int \tan x \, dx = In|\sec x| + C$

8. $\displaystyle\int \cot s \, dx = In|\sin x| + C$

9. $\displaystyle\int \sec x \, dx = In|\sec x + \tan x| + C$

10. $\displaystyle\int \csc x \, dx = In|\csc x - \cot x| + C$

Most of the following problems are average. A few are somewhat challenging. Make careful and precise use of the differential notation dx and dθ and be careful when arithmetically and algebraically simplifying expressions. You should be proficient integrating various powers and rational functions involving trig functions. You may need to use the following additional well-known trig identities.

$$\sin 2x = 2 \sin x \cos x$$

$$\cos^2 x = 2\cos^2 x - 1 \text{ so that } \cos^2 x = \frac{1}{2}\left(1 + \cos^2 2x\right)$$

$$\cos^2 x = 1 - 2\sin^2 x \text{ so that } \sin^2 x = \frac{1}{2}\left(1 - \cos 2x\right)$$

$$\cos^2 x = \cos^2 x - \sin^2 x$$

$$1 + \cot^2 x = \csc^2 x \text{ so that } \cot^2 x = \csc^2 x - 1$$

Problem: Integrate $\int \sqrt{1-x^2}\,dx$

Solution: To integrate $\int \sqrt{1-x^2}\,dx$ use the trig substitution: $X = \sin\theta$

so that, $dx = \cos\theta\,d\theta$

Substitute into the original problem, replacing all forms of x, getting:

$$\int \sqrt{1-x^2}\,dx = \int \sqrt{1-\sin^2\theta}\,\cos\theta\,d\theta$$

$$= \int \sqrt{\cos^2\theta}\,\cos\theta\,d\theta$$

$$= \int \cos\theta\cos\theta\,d\theta$$

$$= \int \cos^2\theta\,du$$

(Recall that $\cos^2\theta = 2\cos^2\theta - 1$ so that $\cos^2\theta = \frac{1}{2}\left(1 + \cos 2\theta\right)$

$$= \int \frac{1}{2}\left(1 + \cos 2\theta\right)du$$

$$= \frac{1}{2}\int \left(1 + \cos 2\theta\right)du$$

$$= \frac{1}{2}\left(\theta + \frac{1}{2}\sin 2\theta\right) + C$$

Recall that $\sin 2\theta = 2\sin\theta\cos\theta$.

$$= \frac{1}{2}\left(\theta + \frac{1}{2}2\sin\theta\cos\theta\right) + C$$

$$= \frac{1}{2}\left(\theta + \sin\theta\cos\theta\right) + C$$

Since,

$$\sin\theta = x = \frac{x}{1} = \frac{opposite}{hypotenuse}$$

We need to write our final answer in terms of x.

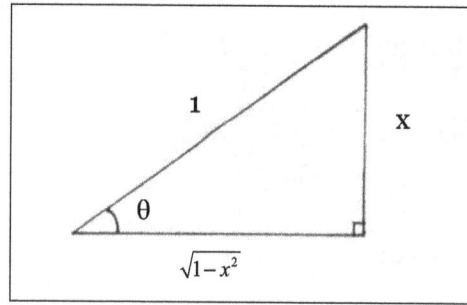

It follows that θ = arcsin x, and from the Pythagorean Theorem that,

$$(adjacent)^2 + (opposite)^2 = (hypotenuse)^2 \rightarrow$$

$$(adjacent)^2 + (x)^2 = (1)^2 \rightarrow adjacent = \sqrt{1-x^2} \rightarrow$$

$$\cos\theta = \frac{adjacent}{hypotenuse} = \frac{\sqrt{1-x^2}}{1} = \sqrt{1-x^2}\,)$$

$$= \frac{1}{2}\left(\arcsin x + x\sqrt{1-x^2}\right) + C$$

Problem: Integrate $\int \dfrac{\left(x^2-1\right)^{3/2}}{x}\,dx$.

Solution: To integrate $\int \dfrac{\left(x^2-1\right)^{3/2}}{x}\,dx$ use the trig substitution,

$$X = \sec\theta$$

so that,

$$dx = \sec\theta\tan\theta\,d\theta$$

Substitute into the original problem, replacing all forms of x, getting:

$$\int\frac{\left(x^2-1\right)^{3/2}}{x}\,dx = \int\frac{\left(\sec^2\theta-1\right)^{3/2}}{\sec\theta}\sec\theta\tan\theta\,d\theta$$

$$= \int\frac{\left(\tan^2\theta\right)^{3/2}}{\sec\theta}\sec\theta\tan\theta\,d\theta$$

$$= \int\frac{\tan^3\theta}{\sec\theta}\sec\theta\tan\theta\,d\theta$$

$$= \int\tan^4\theta\,d\theta$$

$$= \int\tan^2\theta\tan^2\theta\,d\theta$$

Recall that $\tan^2 \theta = \sec^2 \theta - 1$:

$$= \int \tan^2 \theta \left(\sec^2 \theta - 1\right) \ \theta$$

$$= \int \left(\tan^2 \theta \sec^2 \theta - \tan^2 \theta\right) \ \theta$$

$$= \int \left(\tan^2 \theta \sec^2 \theta - \left(\sec^2 \theta - 1\right)\right) \ \theta$$

$$= \int \left(\tan^2 \theta \sec^2 \theta - \sec^2 \theta + 1\right) \ \theta$$

$$= \int \tan^2 \theta \sec^2 \theta \, d\theta - \tan - \theta + \theta + C$$

Now let u = tanθ → du = sec² θ dθ:

$$= \int u^2 \, du - \tan \theta + \theta + C$$

$$= \frac{u^3}{3} - \tan \theta + \theta + C$$

$$= \frac{\left(\tan \theta\right)^3}{3} - \tan \theta + \theta + C$$

We need to write our final answer in terms of x.

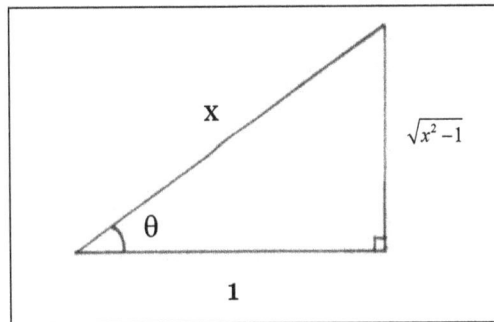

Since x = sec θ it follows that θ = arc sec x and

$$\sec \theta = \frac{x}{1} = \frac{hypotenuse}{adjacent}$$

and from the Pythagorean Theorem that,

$$(adjacent)^2 + (opposite)^2 = (hypotenuse)^2 \ \rightarrow$$

$$(1)^2 + (opposite)^2 = (x)^2 \ \rightarrow \ opposite = \sqrt{x^2 - 1} \ \rightarrow$$

$$\tan \theta = \frac{opposite}{adjacent} = \frac{\sqrt{x^2 - 1}}{1})$$

$$= \frac{1}{3}\left(\sqrt{x^2 - 1}\right)^3 - \sqrt{x^2 - 1} + arc \sec x + C$$

$$= \frac{1}{3}\left(x^2 - 1\right)^{3/2} - \sqrt{x^2 - 1} + arc \sec x + C$$

Problem: Integrate $\int \dfrac{1}{\left(1-x^2\right)^{3/2}}\,dx$.

Solution: To integrate $\int \dfrac{1}{\left(1-x^2\right)^{3/2}}\,dx$ use the trig substitution,

$$X = \sin\theta$$

so that,

$$dx = \cos\theta\,d\theta$$

Substitute into the original problem, replacing all forms of x, getting:

$$\int \frac{1}{\left(1-x^2\right)^{3/2}}\,dx = \int \frac{1}{\left(1-\sin^2\theta\right)^{3/2}}\cos\theta\,d\theta$$

$$= \int \frac{1}{\left(\cos^2\theta\right)^{3/2}}\cos\theta\,d\theta$$

$$= \int \frac{1}{\cos^3\theta}\cos\theta\,d\theta$$

$$= \int \frac{1}{\cos^2\theta}\,d\theta$$

$$= \int \sec^2\theta\,d\theta$$

$$= \tan\theta + C$$

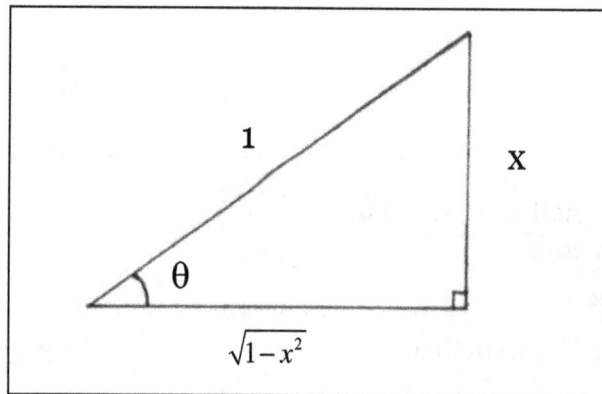

Since $x = \sin\theta$ it follows that,

$$\sin\theta = \frac{x}{1} = \frac{opposite}{hypotenuse}$$

and from the Pythagorean Theorem that,

$$(adjacent)^2 + (opposite)^2 = (hypotenuse)^2 \;\rightarrow$$

$$(adjacent)^2 + (x)^2 = (1)^2 \rightarrow \quad adjacent = \sqrt{1-x^2} \rightarrow$$

$$\tan\theta = \frac{opposite}{adjacent} = \frac{x}{\sqrt{1-x^2}}$$

$$= \frac{x}{\sqrt{1-x^2}} + C$$

Problem: Integrate $\int \frac{\sqrt{x^2+1}}{x} dx$.

Solution: To integrate $\int \frac{\sqrt{x^2+1}}{x} dx$ use the trig substitution,

$$X = \tan\theta$$

so that,

$$dx = \sec^2\theta \, d\theta$$

Substitute into the original problem, replacing all forms of x, getting:

$$\int \frac{\sqrt{x^2+1}}{x} dx = \int \frac{\sqrt{\tan^2\theta+1}}{\tan\theta} \sec^2\theta d\theta$$

$$= \int \frac{\sqrt{\sec^2\theta}}{\tan\theta} \sec^2\theta \, d\theta$$

$$= \int \frac{\sec\theta}{\tan\theta} \sec^2\theta \, d\theta$$

$$= \int \frac{\sec^3\theta}{\tan\theta} d\theta$$

$$= \int \frac{\sec\theta \sec^2\theta}{\tan\theta} d\theta$$

$$= \int \frac{\sec\theta(1+\tan^2\theta)}{\tan\theta} d\theta$$

$$= \int \frac{\cos\theta}{\sin\theta} \cdot \frac{1}{\cos\theta} \cdot (1+\tan^2\theta) d\theta$$

$$= \int \frac{1+\tan^2\theta}{\sin\theta} d\theta$$

$$= \int \frac{1}{\sin\theta}\left(1+\left(\frac{\sin\theta}{\cos\theta}\right)^2\right) d\theta$$

$$= \int \frac{1}{\sin\theta}\left(1+\frac{\sin^2\theta}{\cos^2\theta}\right)d\theta$$

$$= \int \left(\frac{1}{\sin\theta}+\frac{\sin\theta}{\cos^2\theta}\right)d\theta$$

$$= \int \left(\csc\theta+\frac{1}{\cos\theta}\cdot\frac{\sin\theta}{\cos\theta}\right)d\theta$$

$$= \int (\csc\theta+\sec\theta\tan\theta)\,d\theta$$

$$\int \csc\theta\,d\theta = In|\csc\theta-\cot\theta|+C \text{ and } \int \sec\theta\tan\theta\,d\theta = \sec\theta+C.$$

$$= In|\csc\theta-\cot\theta|+\sec\theta+C$$

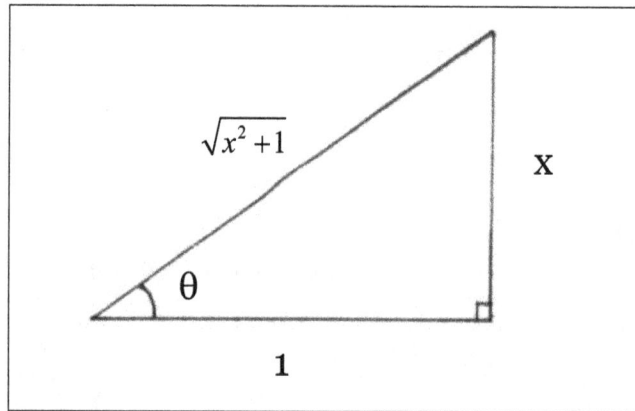

Since $x = \tan\theta$ it follows that,

$$\tan\theta = \frac{x}{1} = \frac{opposite}{adjacent}$$

and from the Pythagorean Theorem that,

$$(adjacent)^2 + (opposite)^2 = (hypotenuse)^2 \rightarrow$$

$$(1)^2 + (x)^2 = (hypotenuse)^2 \rightarrow hypotenuse = \sqrt{x^2+1} \rightarrow$$

$$\sec\theta = \frac{hypotenuse}{adjacent} = \frac{\sqrt{x^2+1}}{1},$$

$$\csc\theta = \frac{hypotenuse}{opposite} = \frac{\sqrt{x^2+1}}{x},$$

and,

$$\cot\theta = \frac{hypotenuse}{opposite} = \frac{1}{x}.)$$

$$= ln\left|\frac{\sqrt{x^2+1}}{x} - \frac{1}{x}\right| + \sqrt{x^2+1} + C$$

Problem: Integrate $\int x^3 \sqrt{4-9x^2}\,dx$.

Solution: To integrate $\int x^3 \sqrt{4-9x^2}\,dx = \int x^3 \sqrt{4\left(1-(9/4)x^2\right)}\,dx = 2\int x^3 \sqrt{1-\left((3/2)x\right)^2}\,dx$ use the trig substitution,

$$X = (2/3)\sin\theta$$

so that,

$$dx = (2/3)\cos\theta\,d\theta$$

Substitute into the original problem, replacing all forms of x, getting:

$$2\int x^3 \sqrt{1-\left((3/2)x\right)^2}\,dx = 2\int \left((2/3)\sin\theta\right)^3 \sqrt{1-\left((3/2)((2/3)\sin^2\theta)\right)^2} (2/3)\cos\theta\,d\theta$$

$$= 4/3\int (2/3)^3 \sin^3\theta\sqrt{1-\sin^2\theta}\,\cos\theta\,d\theta$$

$$= 32/81\int \sin^3\theta\cos\theta\cos\theta\,d\theta$$

$$= 32/81\int \sin^3\theta\cos^2\theta\,d\theta$$

Recall that $\sin^2\theta = 1 - \cos^2\theta$.

$$= 32/81\int \sin\theta\sin^2\theta\cos\theta\,d\theta$$

$$= -32/81\int \sin\theta\left(1-\cos^2\theta\right)\cos^2\theta\,d\theta$$

$$= 32/81\int \sin\theta\left(\cos^2\theta-\cos^4\theta\right)d\theta$$

Now let $u = \cos\theta \rightarrow du = -\sin\theta \rightarrow -du = \sin\theta$.

$$= -32/81\int \left(u^2-u^4\right)d\theta$$

$$= -\frac{32}{81}\left(\frac{u^3}{3} - \frac{u^5}{5}\right) + C$$

$$= -\frac{32}{81}\left(\frac{(\cos\theta)^3}{3} - \frac{(\cos\theta)^5}{\theta}\right) + C$$

Since $x = (2/3)\sin\theta$ it follows that:

$$\sin\theta = \frac{3x}{2} = \frac{opposite}{hypotenuse}$$

We need to write our final answer in terms of x.

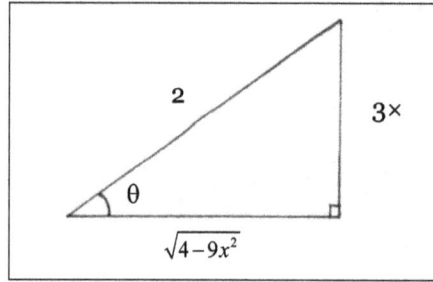

And from the Pythagorean Theorem that,

$$(adjacent)^2 + (opposite)^2 = (hypotenuse)^2 \rightarrow$$

$$(adjacent)^2 + (3x)^2 = (2)^2 \rightarrow adjacent = \sqrt{4-9x^2} \rightarrow$$

$$\cos\theta = \frac{adjacent}{hypotenuse} = \frac{\sqrt{4-9x^2}}{2}$$

$$= -\frac{32}{81}\left(\left(\frac{1}{3}\left(\frac{\sqrt{4-9x^2}}{2}\right)^3 - \frac{1}{5}\left(\frac{\sqrt{4-9x^2}}{2}\right)^5\right)\right) + C$$

$$= -\frac{32}{81}\left(\frac{1}{3}\cdot\frac{1}{8}\left(4-9x^2\right)^{3/2} - \frac{1}{5}\cdot\frac{1}{32}\left(4-9x^2\right)^{5/2}\right) + C$$

$$= -\frac{4}{243}\left(4-9x^2\right)^{3/2} + \frac{1}{405}\left(4-9x^2\right)^{5/2} + C$$

Integration by Partial Fraction

Partial fractions is the name given to a technique of integration that may be used to integrate any ratio of polynomials. A ratio of polynomials is called a rational function. Suppose that N(x) and

D(x) are polynomials. The basic strategy is to write $\frac{N(x)}{D(x)}$ as a sum of very simple, easy to integrate rational functions, namely:

- Polynomials (which are needed only if the degree[1] of N(x) is equal to or strictly bigger than the degree of D(x)).

- Rational functions of the particularly simple form $\frac{A}{(ax+b)^n}$.

- Rational functions of the form $\frac{Ax+B}{(ax^2+bx+c)^m}$.

That is, to find the integral of the rational function on the far right hand side of:

$$x + \frac{1}{x+1} + \frac{1}{x-1} = \frac{x(x+1)(x-1)+(x-1)+(x+1)}{(x+1)(x-1)} = \frac{x^3+x}{x^2-1}$$

You rewrite it as the left hand side and then integrate x and $\dfrac{1}{x+1}$ and $\dfrac{1}{x-1}$ So the main problem is to write a complicated rational function as a sum of simple pieces.

- The denominators on the left hand side are the factors of the denominator $x^2 - 1 = (x - 1)(x + 1)$.

- Use P(x) to denote the polynomial on the left hand side (i.e. P(x) = x) and N(x) to denote the numerator of the right hand side (i.e. N(x) = x³ + x) and D(x) to denote the denominator of the right hand side (i.e. D(x) = x² − 1). Then highest degree term in N(x) is x³. It came from multiplying P(x) by D(x). In particular the degree of N(x) is the sum of the degree P(x) and the degree of D(x). The presence of a polynomial on the left hand side is signalled on the right hand side by the fact that the degree of the numerator is at least as large as the degree of the denominator.

Example: $\displaystyle\int\left(\dfrac{x-3}{x^2-3x+2}\,dx\right)$

In this example, we integrate $\dfrac{N(x)}{D(x)} = \dfrac{x-3}{x^2-3x+2}$.

Step 1: We first check to see if the degree of the numerator, N(x), is strictly smaller than the degree of the denominator D(x). In this example, the numerator, x − 3, has degree one and that is indeed strictly smaller than the degree of the denominator, x² − 3x + 2, which is two. In this case, the first step is not needed and we move on to step 2.

Step 2: The second step is to factor the denominator,

$$x^2 - 3x + 2 = (x - 1)(x - 2)$$

Step 3: The third step is to write $\dfrac{x-3}{x^2-3x-2}$ in the form,

$$\dfrac{x-3}{x^2-3x+2} = \dfrac{A}{x-1} + \dfrac{B}{x-2}$$

For some constants A and B. To determine the values of the constants A, B, we put the right hand side back over the common denominator (x − 1)(x − 2).

$$\dfrac{x-3}{x^2-3x+2} = \dfrac{A}{x-1} + \dfrac{B}{x-2} = \dfrac{A(x-2)+B(x-1)}{(x-1)(x-2)}$$

The fraction on the far left is the same as the fraction on the far right if and only if their numerators are the same,

$$x - 3 = A(x - 2) + B(x - 1)$$

There is a couple of different ways to determine the values of A and B from this equation.

The conceptually clearest procedure is to write the right hand side as a polynomial in standard form (i.e. collect up all x terms and all constant terms):

$$x - 3 = (A + B)x + (-2A - B)$$

For these two polynomials to be the same, the coefficient of x on the left hand side and the coefficient of x on the right hand side must be the same. Similarly the coefficients of x_0 (i.e. the constant terms) must match. This gives us a system of two equations.

$$A + B = 1 \quad -2A - B = -3$$

in the two unknowns A,B. We can solve this system by using the first equation, namely A + B = 1, to determine A in terms of B: A = 1–B. Substituting this into the remaining equation eliminates the A from second equation, leaving one equation in the one unknown B.

$$A = 1 - B \qquad -2A - B = -3$$
$$\Rightarrow \qquad -2(1 - B) - B = -3$$
$$\Rightarrow \qquad B = -1 \; A = 1 - B = 1 - (-1) = 2$$

There is also a second, more efficient, procedure for determining A and B from:

$$x - 3 = A(x - 2) + B(x - 1).$$

This equation must be true for all values of x. In particular, it must be true for x = 1. When x = 1, the factor (x – 1) multiplying B is exactly zero. So B disappears from the equation, leaving us with an easy equation to solve for A:

$$x-3\big|_{x=1} = A(x-2)\big|_{x=1} + B(x-1)\big|_{x=1} \Rightarrow -2 = -A \Rightarrow A = 2.$$

Similarly, when x = 2, the factor (x – 2) multiplying A is exactly zero. So A disappears from the equation, leaving us with an easy equation to solve for B:

$$x-3\big|_{x=2} = A(x-2)\big|_{x=2} + B(x-1)\big|_{x=2} \Rightarrow -1 = -B \Rightarrow B = -1.$$

Step 4: The final step is to integrate,

$$\int \frac{x-3}{x^2 - 3x + 2} dx = \int \frac{2}{x-1} dx + \int \frac{-1}{x-2} dx = 2\ln|x-1| - \ln|x-2| + C.$$

References

- Integral-mathematics, science: britannica.com, Retrieved 3 March, 2019
- Indefinite-integral, integral-calculus, mathematics: technologyuk.net, Retrieved 20 May, 2019
- Definite-integral, integral-calculus, mathematics: technologyuk.net, Retrieved 11 April, 2019
- Integration-by-substitution, integral-calculus, mathematics: technologyuk.net, Retrieved 27 August, 2019
- IntPartFracts: ubc.ca, Retrieved 14 June, 2019

Vectors

The elements of vector space are known as vectors. There are a number of important areas of study related to vectors such as vector functions, gradient, curl and divergence. The chapter closely examines these key concepts of related to vectors to provide an extensive understanding of the subject.

A vector is an object that has both a magnitude and a direction. Geometrically, we can picture a vector as a directed line segment, whose length is the magnitude of the vector and with an arrow indicating the direction. The direction of the vector is from its tail to its head.

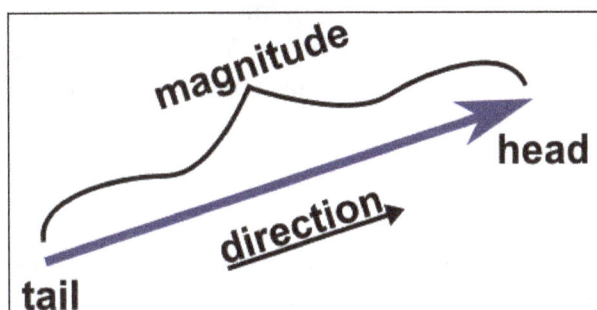

Two vectors are the same if they have the same magnitude and direction. This means that if we take a vector and translate it to a new position (without rotating it), then the vector we obtain at the end of this process is the same vector we had in the beginning.

Two examples of vectors are those that represent force and velocity. Both force and velocity are in a particular direction. The magnitude of the vector would indicate the strength of the force or the speed associated with the velocity.

We denote vectors using boldface as in a or b. Especially when writing by hand where one cannot easily write in boldface, people will sometimes denote vectors using arrows as in $\vec{a}$ or $\vec{b}$, or they use other markings. We won't need to use arrows here. We denote the magnitude of the vector a by ‖a‖. When we want to refer to a number and stress that it is not a vector, we can call the number a scalar. We will denote scalars with italics, as in a or b.

You can explore the concept of the magnitude and direction of a vector using the below figure. Note that moving the vector around doesn't change the vector, as the position of the vector doesn't affect the magnitude or the direction. But if you stretch or turn the vector by moving just its head or its tail, the magnitude or direction will change.

There is one important exception to vectors having a direction. The zero vector, denoted by a boldface 0, is the vector of zero length. Since it has no length, it is not pointing in any particular direction. There is only one vector of zero length, so we can speak of the zero vector.

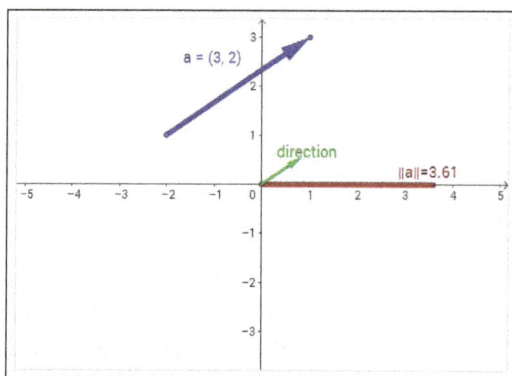

The magnitude and direction of a vector: The blue arrow represents a vector a. The two defining properties of a vector, magnitude and direction, are illustrated by a red bar and a green arrow, respectively. The length of the red bar is the magnitude ‖a‖ of the vector a. The green arrow always has length one, but its direction is the direction of the vector a. The one exception is when a is the zero vector (the only vector with zero magnitude), for which the direction is not defined. You can change either end of a by dragging it with your mouse. You can also move a by dragging the middle of the vector; however, changing the position of the a in this way does not change the vector, as its magnitude and direction remain unchanged.

Operations on Vectors

We can define a number of operations on vectors geometrically without reference to any coordinate system. Here we define addition, subtraction, and multiplication by a scalar.

Addition of Vectors

Given two vectors a and b, we form their sum a + b, as follows. We translate the vector bb until its tail coincides with the head of a. (Recall such translation does not change a vector.) Then, the directed line segment from the tail of a to the head of bb is the vector a +b.

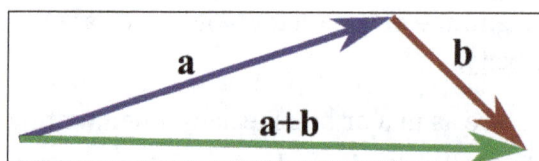

The vector addition is the way forces and velocities combine. For example, if a car is travelling due north at 20 miles per hour and a child in the back seat behind the driver throws an object at 20 miles per hour toward his sibling who is sitting due east of him, then the velocity of the object (relative to the ground!) will be in a north-easterly direction. The velocity vectors form a right triangle, where the total velocity is the hypotenuse. Therefore, the total speed of the object (i.e., the magnitude of the velocity vector) is $\sqrt{20^2 + 20^2} = 20\sqrt{2}$ miles per hour relative to the ground.

Addition of vectors satisfies two important properties:

1. The commutative law, which states the order of addition doesn't matter:

    a + b = b + a.

This law is also called the parallelogram law, as illustrated in the below image. Two of the edges of the parallelogram define a + b, and the other pair of edges define b + a. But, both sums are equal to the same diagonal of the parallelogram.

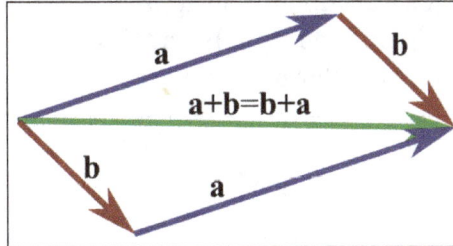

2. The associative law, which states that the sum of three vectors does not depend on which pair of vectors is added first:

(a + b) + c = a + (b + c).

You can explore the properties of vector addition with the following figure.

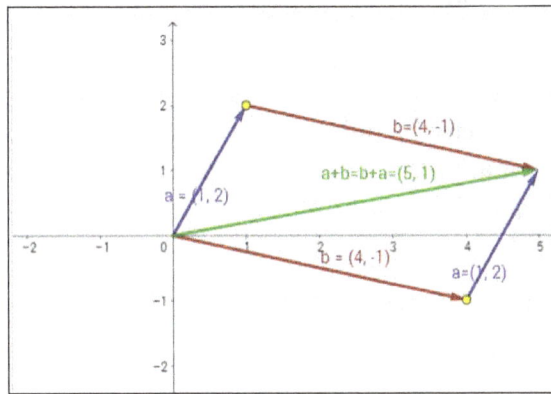

The sum of two vectors: The sum a + b of the vector a (blue arrow) and the vector b (red arrow) is shown by the green arrow. As vectors are independent of their starting position, both blue arrows represent the same vector a and both red arrows represent the same vector b. The sum a + b can be formed by placing the tail of the vector b at the head of the vector a. Equivalently, it can be formed by placing the tail of the vector a at the head of the vector b. Both constructions together form a parallelogram, with the sum a + b being a diagonal. (For this reason, the commutative law a + b = b + a is sometimes called the parallelogram law.) You can change a and b by dragging the yellow points.

Vector Subtraction

Before we define subtraction, we define the vector −a, which is the opposite of a. The vector −a is the vector with the same magnitude as a but that is pointed in the opposite direction.

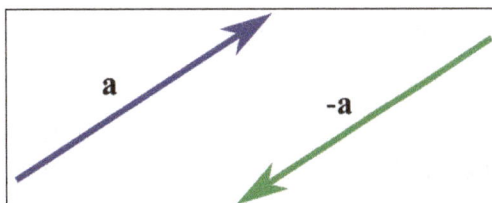

We define subtraction as addition with the opposite of a vector:

$$b - a = b + (-a).$$

This is equivalent to turning vector a around in the applying the above rules for addition. Can you see how the vector x in the below figure is equal to b − a? Notice how this is the same as stating that a + x = b, just like with subtraction of scalar numbers.

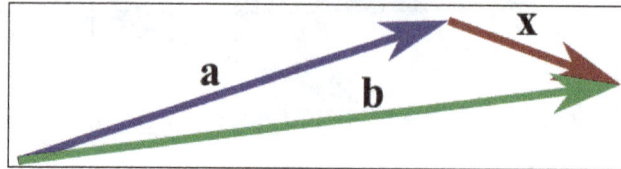

Scalar Multiplication

Given a vector a and a real number (scalar) λ, we can form the vector λa as follows. If λ is positive, then λa is the vector whose direction is the same as the direction of a and whose length is λ times the length of a. In this case, multiplication by λ simply stretches (if $\lambda > 1$) or compresses (if $0 < \lambda < 1$) the vector a.

If, on the other hand, λ is negative, then we have to take the opposite of a before stretching or compressing it. In other words, the vector λa points in the opposite direction of a, and the length of λa is $|\lambda|$ times the length of a. No matter the sign of λ, we observe that the magnitude of λa is $|\lambda|$ times the magnitude of a: $\|\lambda a\| = |\lambda| \|a\|$. Scalar multiplications satisfies many of the same properties as the usual multiplication.

1. $s(a + b) = sa + sb$ (distributive law, form 1)

2. $(s + t) a = sa + ta$ (distributive law, form 2)

3. $1a = a$

4. $(-1) a = -a$

5. $0 a = 0$

In the last formula, the zero on the left is the number 0, while the zero on the right is the vector 0, which is the unique vector whose length is zero.

If $a = \lambda b$ for some scalar λ, then we say that the vectors a and b are parallel.

Vector Functions

A vector function covers a set of multidimensional vectors at the intersection of the domains of f, g, and h.

Vector valued functions, also called vector functions, allow you to express the position of a point in multiple dimensions within a single function. These can be expressed in an infinite number of dimensions, but are most often expressed in two or three. The input into a vector valued function can be a vector or a scalar. In this atom we are going to introduce the properties and uses of the vector valued functions.

Properties of Vector Valued Functions

A vector valued function allows you to represent the position of a particle in one or more dimensions. A three-dimensional vector valued function requires three functions, one for each dimension. In Cartesian form with standard unit vectors (i, j, k), a vector valued function can be represented in either of the following ways,

$$r(t) = f(t)i + g(t)j + h(t)k$$
$$r(t) = \langle f(t), g(t), h(t) \rangle$$

where, t is being used as the variable. This is a three dimensional vector valued function. The graph shows a visual representation of:

$r(t) = \langle 2\cos(t), 4\sin(t), t \rangle.$

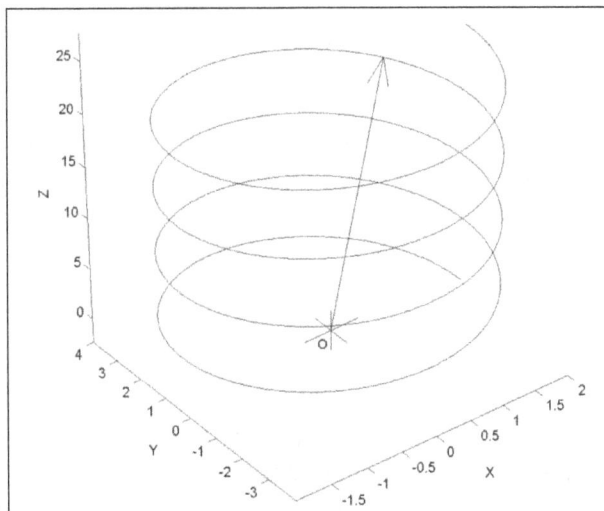

Vector-Valued Function: This graph of a parametric curve (a simple vector-valued function with a single parameter of dimension 11). The graph is of the curve: $\langle 2\cos(t), 4\sin(t), t \rangle$ where t goes from 0 to 8π.

This can be broken down into three separate functions called component functions:

$x(t) = 2\cos(t) y(t) = 4\sin(t) z(t) = t.$

If you were to take a cross section, with the cut perpendicular to any of the three axes, you would see the graph of that function. For example, if you were to slice the three-dimensional shape perpendicular to the z-axis, the graph you would see would be of the function $z(t) = t$. The domain of a vector valued function is a domain that satisfies all of the component functions. It can be found by taking the intersection of the individual component function domains. The vector valued functions can be manipulated in the same way as a vector; they can be added, subtracted, and the dot product and the cross product can be found.

Example: For this example, we will use time as our parameter. The following vector valued function represents time, t,

r(t) = f(t)i + g(t)j + h(t)k

This function is representing a position. Therefore, if we take the derivative of this function, we will get the velocity:

$$\frac{dr(t)}{dt} = f(t)i' + g(t)j' + h(t)k'$$

$$= v(t)$$

If we differentiate a second time, we will be left with acceleration:

$$\frac{dv(t)}{dt} = a(t)$$

Arc Length and Speed

Arc length and speed are, respectively, a function of position and its derivative with respect to time.

Since length is a magnitude that involves position, it is easy to deduce that the derivative of a length, or position, will give you the velocity—also known as speed—of a function. This is because a derivative gives you a rate of change with respect to a parameter. Velocity is the rate of change of a position with respect to time. Let's start this atom by looking at arc length with calculus.

Arc Length

The arc length is the length you would get if you took a curve, straightened it out, and then measured the length of that line. The arc length can be found using geometry, but for the sake of this atom, we are going to use integration. The arc length is approximated by connecting a finite number of points along and curve, connecting those lines to create a string of very small straight lines, and adding them together. To find this using integration, we should start out by using the Pythagorean Theorem for length of the different sides of a triangle,

$$ds^2 = dx^2 + dy^2 \frac{ds^2}{dx^2}$$

$$= 1 + \frac{dy^2}{dx^2} ds$$

$$= \sqrt{1 + \frac{dy^2}{dx^2}} \cdot dx$$

$$= \int_a^b \sqrt{1 + f'(x)^2} \cdot dx$$

where s is the arc length. If $x = X(t)$ and $y = Y(t)$,

$$s = \int_a^b \sqrt{1 + f'(x)^2}\,....dx$$

$$= \int_a^b \sqrt{[X'(t)]^2 + [Y'(t)]}\,....dt$$

$$= \int_a^b \sqrt{dx^2 + dy^2}$$

$$= \int_a^b \sqrt{\frac{dx^2}{dt} + \frac{dy^2}{dt}}\,...dt$$

Since this is a function of position and is defined by x, we need to have a derivative that is in respect to x:

$$\int_a^b \sqrt{1 + \frac{dy^2}{dx}} * dx$$

Obviously some cases require polar coordinates instead of Cartesian. In polar coordinates:

r = f(θ)

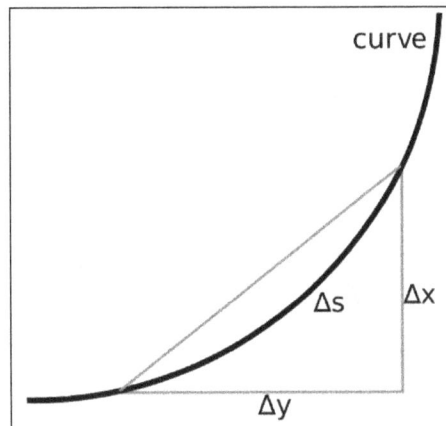

Curves and the Pythagorean Theorem: For a small piece of curve, Δs can be approximated with the Pythagorean theorem.

Arc Speed

Now that the hard part is over, we can easily find the speed along this curve. Since speed is in relation to time and not position, we need to revert back to the arc length with respect to time:

$$\int_a^b \sqrt{\frac{dx^2}{dt} + \frac{dy^2}{dt}}\,.dt$$

Then, differentiate with respect to time:

$$v(t) = s' = \sqrt{[X'(t)]^2 + [Y'(t)]^2}$$

Calculus of Vector-valued Functions

A vector function is a function that can behave as a group of individual vectors and can perform differential and integral operations.

A vector-valued function, also referred to as a vector function, is a mathematical function of one or more variables whose range is a set of multidimensional vectors or infinite-dimensional vectors. The input of a vector-valued function could be a scalar or a vector. The dimension of the domain is not defined by the dimension of the range.

A common example of a vector valued function is one that depends on a single real number parameter t, often representing time, producing a vector v(t) as the result. In terms of the standard unit vectors i, j, *k* of Cartesian 3-space, these specific type of vector-valued functions are given by expressions such as,

$$r(t) = f(t) \, I + g(t) \, j + h \, (t)k$$

where f(t), g(t), and h(t) are the coordinate functions of the parameter t. The vector v(t) has its tail at the origin and its head at the coordinates evaluated by the function.

Vector functions can also be referred to in a different notation:

$$r(t) = \langle f(t), g(t), h(t) \rangle$$

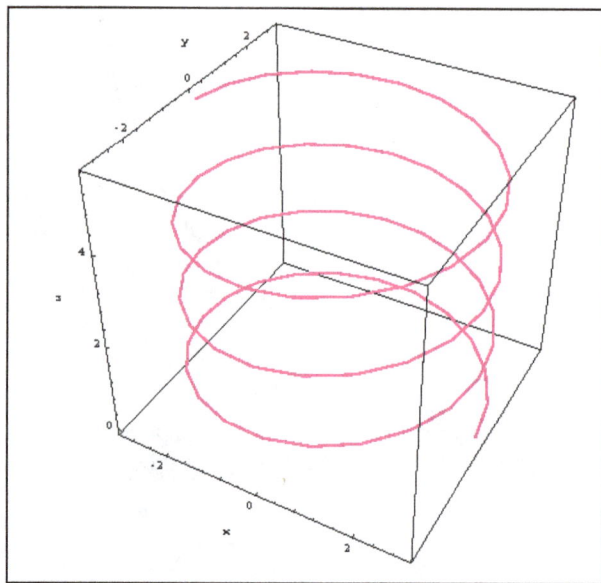

Vector valued function: This graph is a visual representation of the three-dimensional vector-valued function r(t) = ⟨2cos(t), 4sin(t),t⟩. This can be broken down into three separate functions called component functions: x(t) = 2cos (t)y (t) = 4sin(t)z(t) = t.

Vector calculus is a branch of mathematics that covers differentiation and integration of vector fields in any number of dimensions. Because vector functions behave like individual vectors, you can manipulate them the same way you can a vector. Vector calculus is used extensively throughout physics and engineering, mostly with regard to electromagnetic fields, gravitational fields, and fluid flow. When taking the derivative of a vector function, the function should be treated as a group of individual functions.

Vector functions are used in a number of differential operations, such as gradient (measures the rate and direction of change in a scalar field), curl (measures the tendency of the vector function to rotate about a point in a vector field), and divergence (measures the magnitude of a source at a given point in a vector field).

Arc Length and Curvature

The curvature of an object is the degree to which it deviates from being flat and can be found using arc length.

Arc Curvature

The curvature of an arc is a value that represents the direction and sharpness of a curve. On any curve, there is a center of curvature, C. This is the intersection point of two infinitely close normal to this curve. The radius, R, is the distance from this intersection point to the center of curvature.

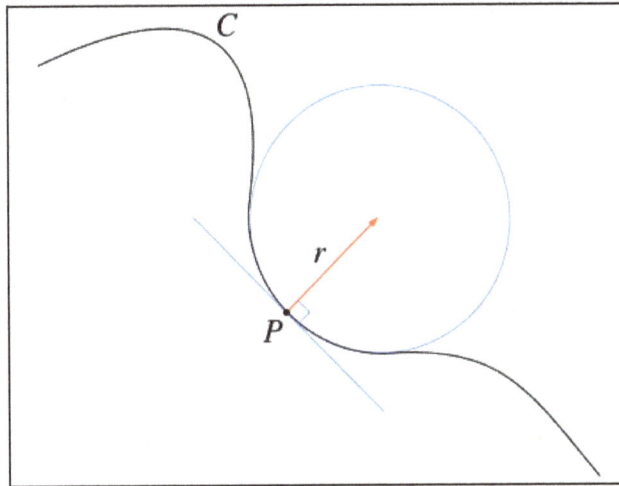

Curvature: Curvature is the amount an object deviates from being flat.

Given any curve C and a point P on it, there is a unique circle or line which most closely approximates the curve near P. The curvature of C at P is then defined to be the curvature of that circle or line. The radius of curvature is defined as the reciprocal of the curvature.

In order to find the value of the curvature, we need to take the parameter time, s, and the unit tangent vector, which in this case is the same as the unit velocity vector, T, which is also a function of time. The curvature is a magnitude of the rate of change of the tangent vector, T,

$$k = \left\| \frac{dT}{ds} \right\|$$

where κ is the curvature and $\frac{dT}{ds}$ is the acceleration vector (the rate of change of the velocity vector over time).

Relation between Curvature and Arc Length

The curvature can also be approximated using limits. Given the points P and Q on the curve, lets call the arc length s(P,Q), and the linear distance from P to Q will be denoted as d(P,Q). The curvature of the arc at point P can be found by obtaining the limit:

$$\kappa(P) = \frac{\lim}{Q \to P} \sqrt{\frac{24*\left(s(P,Q)-d(P,Q)\right)}{s(P,Q)^3}}$$

In order to use this formula, you must first obtain the arc length of the curve from points P to Q and length of the linear segment that connect points P and Q. In Cartesian coordinates:

$$\int_a^b \sqrt{1 + \frac{dy^2}{dx}} * dx \ .$$

Tangent Vectors and Normal Vectors

A vector is normal to another vector if the intersection of the two form a 90-degree angle at the tangent point.

In order for a vector to be normal to an object or vector, it must be perpendicular with the directional vector of the tangent point. The intersection formed by the two objects must be a right angle.

Normal Vectors

An object is normal to another object if it is perpendicular to the point of reference. That means that the intersection of the two objects forms a right angle. Usually, these vectors are denoted as n.

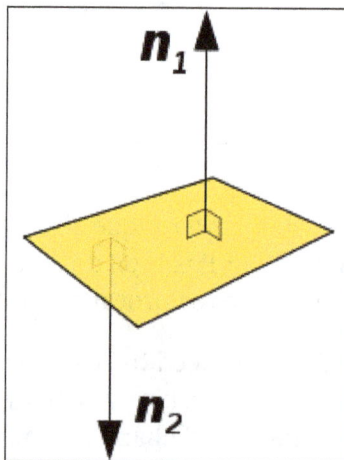

Normal Vector: These vectors are normal to the plane because the intersection between them and the plane makes a right angle.

Not only can vectors be 'normal' to objects, but planes can also be normal.

Dot Product

As we covered in another atom, one of the manipulations of vectors is called the Dot Product. When you take the dot product of two vectors, your answer is in the form of a single value, not a vector. In order for two vectors to be normal to each other, the dot product has to be zero.

$$a \cdot b = 0$$
$$= a_1 b_1 + a_2 b_2 + a_3 b_3$$
$$= |a||b|\cos\theta$$

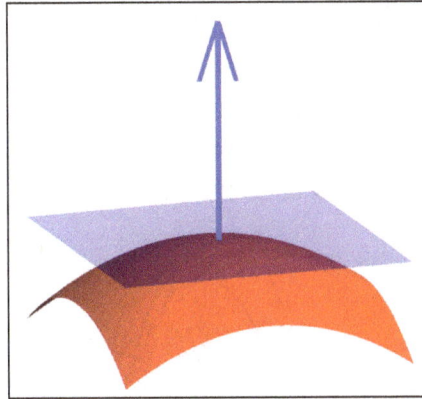

Normal Plane: A plane can be determined as normal to the object if the directional vector of the plane makes a right angle with the object at its tangent point. This plane is normal to the point on the sphere to which it is tangent. Each point on the sphere will have a unique normal plane.

Tangent Vectors

Tangent vectors are almost exactly like normal vectors, except they are tangent instead of normal to the other vector or object. These vectors can be found by obtaining the derivative of the reference vector, r(t):

r(t) = f(t) i + g (t)j + h (t)k

Gradient

The gradient of a function w = f(x, y, z) is the vector function:

$$\nabla f = grad\ f = < \frac{\partial f}{\partial x}(x,y,z), \frac{\partial f}{\partial y}(x,y,z), \frac{\partial f}{\partial z}(x,y,z) >$$

For a function of two variables z = f(x, y), the gradient is the two-dimensional vector < f_x (x, y), f_y (x, y)>. This definition generalizes in a natural way to functions of more than three variables.

Examples: For the function z = f(x, y) = 4x 2 + y 2. The gradient is,

$$grad\ f = <8x, 2y>.$$

For the function w = g(x, y, z) = exp(x y z) + sin(xy), the gradient is:

$$grad\ g = < yze^{xyz} + y\cos(xy), xze^{xyz} + x\cos(xy), xye^{xyz} >$$

Geometric Description of the Gradient Vector

There is a nice way to describe the gradient geometrically. Consider z = f(x, y) = 4x² + y². The surface defined by this function is an elliptical paraboloid. This is a bowl-shaped surface. The bottom of the bowl lies at the origin. The figure below shows the level curves, defined by f(x, y) = c, of the surface. The level curves are the ellipses 4x² + y²= c.

The gradient vector <8x, 2y> is plotted at the 3 points (sqrt(1.25), 0), (1, 1), (0, sqrt(5)). As the plot shows, the gradient vector at (x, y) is normal to the level curve through (x, y). the gradient vector points in the direction of greatest rate of increase of f(x, y).

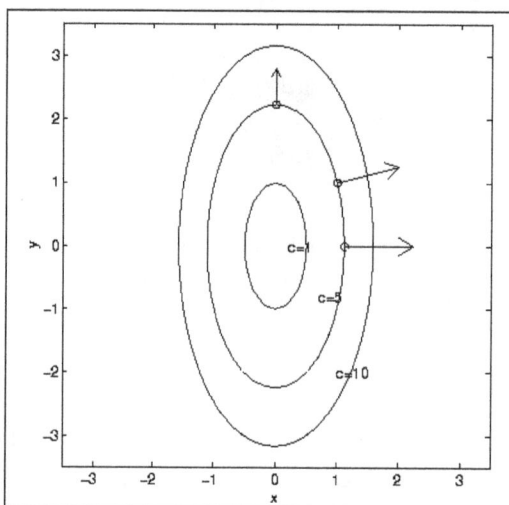

In three dimensions the level curves are level surfaces. Again, the gradient vector at (x, y, z) is normal to level surface through (x, y, z).

Directional Derivatives

For a function z = f(x, y), the partial derivative with respect to x gives the rate of change of f in the x direction and the partial derivative with respect to y gives the rate of change of f in the y direction. How do we compute the rate of change of f in an arbitrary direction?

The rate of change of a function of several variables in the direction u is called the directional derivative in the direction u. Here u is assumed to be a unit vector. Assuming w = f(x, y, z) and u = <u_1, u_2, u_3 > we have:

$$D_u f = grad\ f \cdot u = \frac{\partial f}{\partial x} u_1 + \frac{\partial f}{\partial y} u_2 + \frac{\partial f}{\partial z} u_3.$$

Hence, the directional derivative is the dot product of the gradient and the vector u. Note that if u is a unit vector in the x direction, u = <1, 0, 0>, then the directional derivative is simply the partial derivative with respect to x. For a general direction, the directional derivative is a combination of the all three partial derivatives.

Example: What is the directional derivative in the direction <1, 2> of the function z = f(x, y)= 4x² + y² at the point x = 1 and y = 1. The gradient is <8x, 2y>, which is <8, 2> at the point x = 1 and y = 1.

The direction u is <2, 1>. Converting this to a unit vector, we have <2, 1>/sqrt(5). Hence:

$$Duf = grad\ f \cdot u = <8,2> \cdot \frac{2}{\sqrt{5}}, \frac{1}{\sqrt{5}} >= \frac{18}{\sqrt{5}}.$$

Directions of Greatest Increase and Decrease

The directional derivative can also be written:

$$Du\ f = grad\ f \cdot u = |grad\ f||u|\cos\theta$$

where theta is the angle between the gradient vector and u. The directional derivative takes on its greatest positive value if theta = 0. Hence, the direction of greatest increase of f is the same direction as the gradient vector. The directional derivative takes on its greatest negative value if theta = pi (or 180 degrees). Hence, the direction of greatest decrease of f is the direction opposite to the gradient vector.

Curl

The curl of a vector field, denoted curl(F) or del ∇ x F (the notation used in this work), is defined as the vector field having magnitude equal to the maximum "circulation" at each point and to be oriented perpendicularly to this plane of circulation for each point. More precisely, the magnitude of ∇x F is the limiting value of circulation per unit area. Written explicitly,

$$(\nabla \times F) \cdot \hat{n} = \lim_{A \to 0} \frac{\oint_C F \cdot ds}{A}$$

where the right side is a line integral around an infinitesimal region of area Â that is allowed to shrink to zero via a limiting process and $\hat{n}$ is the unit normal vector to this region. If $\nabla \times F = 0$, then the field is said to be an irrotational field. The symbol ∇ is variously known as "nabla" or "del."

The physical significance of the curl of a vector field is the amount of "rotation" or angular momentum of the contents of given region of space. It arises in fluid mechanics and elasticity theory. It is also fundamental in the theory of electromagnetism, where it arises in two of the four Maxwell equations,

$$\nabla \times E = \frac{\partial B}{\partial t}$$

$$\nabla \times B = \mu_0 J + \varepsilon_0 \mu_0 \frac{\partial E}{\partial t},$$

where MKS units have been used here, E denotes the electric field, B is the magnetic field, μ_0 is a constant of proportionality known as the permeability of free space, J is the current density, and ε_0 is another constant of proportionality known as the permittivity of free space. Together with the two other of the Maxwell equations, these formulas describe virtually all classical and relativistic properties of electromagnetism.

In Cartesian coordinates, the curl is defined by:

$$\nabla \times F = \left(\frac{\partial F_z}{\partial y} - \frac{\partial F_y}{\partial_z} \right) \hat{x} + \left(\frac{\partial F_x}{\partial z} - \frac{\partial F_z}{\partial x} \right) \hat{y} + \left(\frac{\partial F_y}{\partial x} - \frac{\partial F_x}{\partial y} \right) \hat{z}.$$

This provides the motivation behind the adoption of the symbol $\nabla \times$ for the curl, since interpreting ∇ as the gradient operator $\nabla = \left(\partial / \partial x, \partial / \partial y, \partial / \partial z \right)$, the "cross product" of the gradient operator with F is given by,

$$\nabla \times F = \begin{vmatrix} \hat{x} & \hat{y} & \hat{z} \\ \dfrac{\partial}{\partial x} & \dfrac{\partial}{\partial y} & \dfrac{\partial}{\partial z} \\ F_x & F_y & F_z \end{vmatrix}$$

which is precisely equation $\nabla \times F = \left(\dfrac{\partial F_z}{\partial y} - \dfrac{\partial F_y}{\partial_z} \right) \hat{x} + \left(\dfrac{\partial F_x}{\partial z} - \dfrac{\partial F_z}{\partial x} \right) \hat{y} + \left(\dfrac{\partial F_y}{\partial x} - \dfrac{\partial F_x}{\partial y} \right) \hat{z}.$ A somewhat

more elegant formulation of the curl is given by the matrix operator equation:

$$\nabla \times F = \begin{vmatrix} 0 & -\dfrac{\partial}{\partial z} & \dfrac{\partial}{\partial y} \\ \dfrac{\partial}{\partial z} & 0 & -\dfrac{\partial}{\partial x} \\ -\dfrac{\partial}{\partial y} & \dfrac{\partial}{\partial x} & 0 \end{vmatrix} F$$

The curl can be similarly defined in arbitrary orthogonal curvilinear coordinates using,

$$F \equiv F_1 \hat{u}_1 + F_2 \hat{u}_2 + F_3 \hat{u}_3$$

and defining,

$$h_i \equiv \left| \frac{\partial r}{\partial u_i} \right|,$$

as,

$$\nabla \times F \equiv \frac{1}{h_1 h_2 h_3} \begin{vmatrix} h_1 \hat{u}_1 & h_2 \hat{u}_2 & h_3 \hat{u}_3 \\ \dfrac{\partial}{\partial u_1} & \dfrac{\partial}{\partial u_2} & \dfrac{\partial}{\partial u3} \\ h_1 F_1 & h_2 F_2 & h_3 F_3 \end{vmatrix}$$

$$\frac{1}{h_2 h_3} \left[\frac{\partial}{\partial u_2} (h_3 F_3) - \frac{\partial}{\partial u_3} (h_2 F_2) \right] \hat{u}_1 + \frac{1}{h_1 h_3}$$

$$\left[\frac{\partial}{\partial u_3}\left(h_1 F_1\right)-\frac{\partial}{\partial u_1}\left(h_3 F_3\right)\right]\hat{u}_2 + \frac{1}{h_1 h_2}\left[\frac{\partial}{\partial u_1}\left(h_2 F_2\right)-\frac{\partial}{\partial u_2}\left(h_1 F_1\right)\right]\hat{u}_3.$$

The curl can be generalized from a vector field to a tensor field as,

$$\left(\nabla \times A\right)^\alpha = \epsilon^{\alpha v}\, A_{v,u,}$$

where ε_{ijk} is the permutation tensor and ";" denotes a covariant derivative.

Divergence

Divergence measures the change in density of a fluid flowing according to a given vector field.

Notation and Formula for Divergence

The notation for divergence uses the same symbol "∇". As with the gradient, we think of this symbol loosely as representing a vector of partial derivative symbols.

$$\nabla = \begin{bmatrix} \dfrac{\partial}{\partial x} \\[2mm] \dfrac{\partial}{\partial y} \\[2mm] \vdots \end{bmatrix}$$

We write the divergence of a vector-valued function $\vec{v}$ (x, y, ...), with, vector, on top, left parenthesis, x, comma, y, comma, dots, space, right parenthesis like this:

$$\nabla \cdot \vec{v} \leftarrow \text{Divergence } of\ \vec{v}$$

This is mildly nonsensical since ∇ isn't really a vector. Its entries are operators, not numbers. Nevertheless, using this dot product notation is super helpful for remembering how to compute divergence, just take a look:

$$\nabla \cdot \vec{v} = \begin{bmatrix} \dfrac{\partial}{\partial x} \\[2mm] \dfrac{\partial}{\partial y} \end{bmatrix} \cdot \begin{bmatrix} 2x - y \\ y^2 \end{bmatrix}$$

$$= \frac{\partial}{\partial x}\left(2x - y\right) + \frac{\partial}{\partial y}\left(y^2\right)$$

$$= 2 + 2y$$

More generally, the divergence can apply to vector-fields of any dimension. This means $\vec{v}$, with, vector, on top can have any number of input variables, as long as its output has the same dimensions. Otherwise, it couldn't represent a vector field. If we write $\vec{v}$, with, vector, on top component-wise like this:

$$\vec{v}(x_1,...,x_n) = \begin{bmatrix} v_1(x_1,...,x_n) \\ \vdots \\ v_n(x_1,...,x_n) \end{bmatrix}$$

Then the divergence of $\vec{v}$, on top looks like this:

$$\nabla \cdot \vec{v} = \begin{bmatrix} \dfrac{\partial}{\partial x_1} \\ \vdots \\ \dfrac{\partial}{\partial x_n} \end{bmatrix} \cdot \begin{bmatrix} v_1 \\ \vdots \\ v_n \end{bmatrix} = \frac{\partial v_1}{\partial x_1} + ... + \frac{\partial v_n}{\partial x_n}$$

Let's summarize this with a quick diagram:

A useful mnemonic is to imagine taking this dot product. This is the reason for the "∇" notation.

Interpretation of Divergence

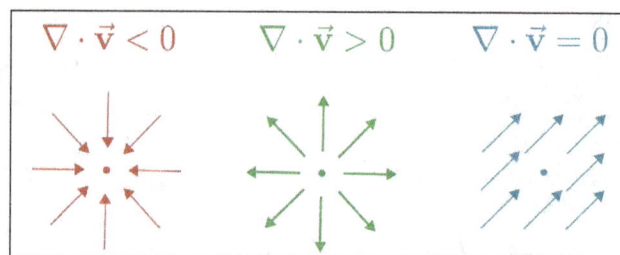

Let's say you evaluate the divergence of a function $\vec{v}$ at some point (x_0, y_0) and it comes out negative.

$$\nabla \cdot \vec{v}(x_0, y_0) < 0$$

This means a fluid flowing along the vector field defined by $\vec{v}$ would tend to become more Dense at the point (x_0, y_0). For example, the following animation shows a vector field with negative divergence at the origin.

On the other hand, if the divergence at a point (x_0, y_0) is positive:

$$\nabla \cdot \vec{v}(x_0, y_0) > 0$$

The fluid flowing along the vector field becomes less dense around (x_0, y_0).

Finally, the concept of zero-divergence is very important in fluid dynamics and electrodynamics. It indicates that even though a fluid flows freely, its density stays constant. This is particularly handy when modeling incompressible fluids, such as water. In fact, the very idea that a fluid is incompressible can be tightly communicated with the following equation:

$$\nabla \cdot \vec{v} = 0$$

Such vector fields are called "divergence-free."

Sources and Sinks

Sometimes, for points with negative divergence, instead of thinking about a fluid getting more dense after a momentary fluid motion, some people imagine the fluid draining at that point while the fluid flows constantly. Here's what this might look like: As such, points of negative divergence are often called "sinks."

Likewise, instead of thinking of points with positive divergence as becoming less dense during a momentary motion, these points might be thought of as "sources" constantly generating more fluid particles.

References

- Vector-introduction: mathinsight.org, Retrieved 2 February, 2019
- Vector-functions, calculus: lumenlearning.com, Retrieved 8 June, 2019
- Curl: wolfram.com, Retrieved 14 January, 2019
- Divergence, divergence-and-curl, math: khanacademy.org, Retrieved 15 March, 2019

Theorems used in Calculus

The field of calculus uses a number of theorems such as Rolle's Theorem, divergence theorem, gradient theorem, Stokes' theorem, Green's theorem and mean value theorem. The diverse applications of these theorems have been thoroughly discussed in this chapter.

Rolle's Theorem

Suppose that a function $f(x)$ is continuous on the closed interval [a, b] and differentiable on the open interval (a, b). Then if f(a) = f(b), then there exists at least one point c in the open interval (a, b) for which $f'(c) = 0$.

Geometric Interpretation

There is a point c on the interval (a, b) where the tangent to the graph of the function is horizontal.

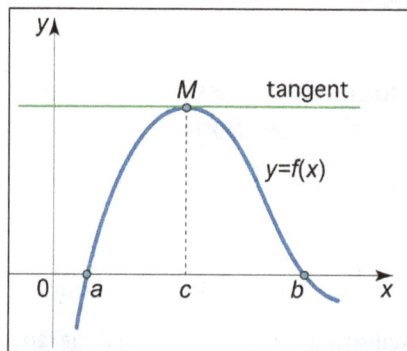

All conditions of Rolle's theorem are necessary for the theorem to be true:

1. $f(x)$ is continuous on the closed interval [a, b];

2. $f(x)$ is differentiable on the open interval (a, b);

3. f(a) = f(b).

Some Counter-examples

Consider $f(x)$ = {x} ({x} is the fractional part function) on the closed interval [0, 1]. The derivative of the function on the open interval (0, 1) is everywhere equal to 1. In this case, the Rolle's theorem fails because the function $f(x)$ has a discontinuity at x = 1 (that is, it is not continuous everywhere on the closed interval [0, 1]).

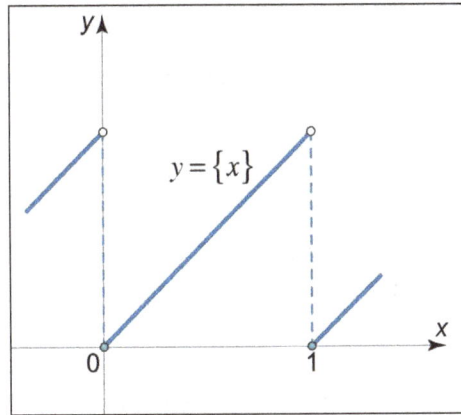

Consider $f(x) = |x|$ (where $|x|$ is the absolute value of x) on the closed interval $[-1, 1]$. This function does not have derivative at $x = 0$. Though $f(x)$ is continuous on the closed interval $[-1, 1]$, there is no point inside the interval $(-1, 1)$ at which the derivative is equal to zero. The Rolle's theorem fails here because $f(x)$ is not differentiable over the whole interval $(-1, 1)$.

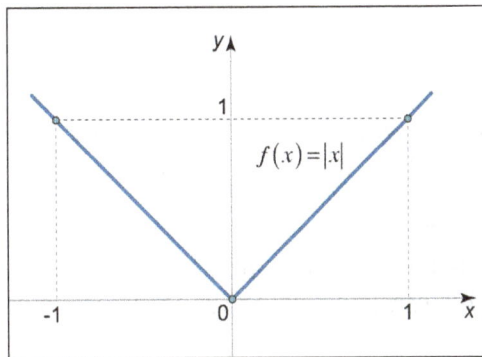

The linear function $f(x) = x$ is continuous on the closed interval $[0,1]$ and differentiable on the open interval $(0, 1)$. The derivative of the function is everywhere equal to 1 on the interval. So the Rolle's theorem fails here. This is explained by the fact that the 3rd condition is not satisfied (since f(0) ≠ f(1).)

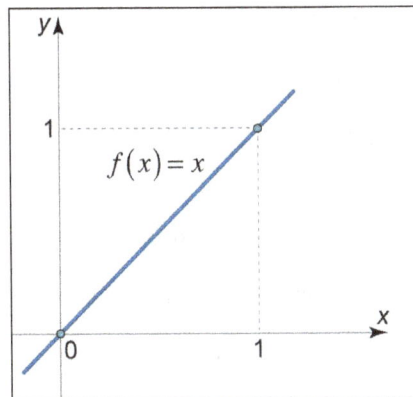

In modern mathematics, the proof of Rolle's theorem is based on two other theorems – the Weierstrass extreme value theorem and Fermat's theorem.

Weierstrass Extreme Value Theorem

If a function $f(x)$ is continuous on a closed interval [a, b], then it attains the least upper and greatest lower bounds on this interval.

Fermat's Theorem

Let a function $f(x)$ be defined in a neighborhood of the point x_0 and differentiable at this point. Then, if the function $f(x)$ has a local extremum at x_0 then:

$$f'(x_0) = 0.$$

Consider now Rolle's theorem in a more rigorous presentation. Let a function $y = f(x)$ be continuous on a closed interval [a, b], differentiable on the open interval (a, b), and takes the same values at the ends of the segment:

$$f(a) = f(b).$$

Then on the interval (a, b) there exists at least one point $c \in (a, b)$, in which the derivative of the function $f(x)$ is zero:

$$f'(c) = 0.$$

Proof: If the function $f(x)$ is constant on the interval [a, b], then the derivative is zero at any point of the interval (a, b), i.e. in this case the statement is true.

If the function $f(x)$ is not constant on the interval [a, b], then by the Weierstrass theorem, it reaches its greatest or least value at some point c of the interval (a, b), i.e. there exists a local extremum at the point c. Then by Fermat's theorem, the derivative at this point is equal to zero:

$$f'(c) = 0.$$

Physical Interpretation

Rolle's theorem has a clear physical meaning. Suppose that a body moves along a straight line, and after a certain period of time returns to the starting point. Then, in this period of time there is a moment, in which the instantaneous velocity of the body is equal to zero.

Example: Let $f(x) = x^2 + 2x$. Find all values of c in the interval $[-2, 0]$ such that $f'(c) = 0$.

Solution: First of all, we need to check that the function $f(x)$ satisfies all the conditions of Rolle's theorem.

1. $f(x)$ is continuous in $[-2,0]$ as a quadratic function;

2. It is differentiable everywhere over the open interval $(-2,0)$;

3. Finally,
$$f(-2) = (-2)^2 + 2 \cdot (-2) = 0,$$
$$f(0) = 0^2 + 2.0 = 0,$$
$$\Rightarrow f(-2) = f(0).$$

So we can use Rolle's theorem.

To find the point c we calculate the derivative,

$$f'(x) = (x^2 + 2x)' = 2x + 2$$

and solve the equation $f'(c) = 0$:

$$f'(c) = 2c + 2 = 0, \Rightarrow c = -1.$$

Thus, $f'(c) = 0$ for $c = -1$.

Example: Given the function $f(x) = x^2 - 6x + 5$. Find all values of c in the open interval $(2, 4)$ such that $f'(c) = 0$.

Solution: First we determine whether Rolle's theorem can be applied to $f(x)$ on the closed interval.

The function is continuous on the closed interval.

The function is differentiable on the open interval. Its derivative is,

The function has equal values at the endpoints of the interval,

$$f(2) = 2^2 - 6 \cdot 2 + 5 = -3,$$
$$f(4) = 4^2 - 6 \cdot 4 + 5 = -3.$$

This means that we can apply Rolle's theorem. Solve the equation to find the point c:

$$f'(c) = 0, \Rightarrow 2c - 6 = 0, \Rightarrow c = 3.$$

Divergence Theorem

The divergence theorem is about closed surfaces, so let's start there. By a closed surface S we will mean a surface consisting of one connected piece which doesn't intersect itself, and which completely encloses a single finite region D of space called its interior. The closed surface S is then said to be the boundary of D; we include S in D. A sphere, cube, and torus (an inflated bicycle inner tube) are all examples of closed surfaces. On the other hand, these are not closed surfaces: a plane, a sphere with one point removed, a tin can whose cross-section looks like a figure (it intersects itself), an infinite cylinder.

A closed surface always has two sides, and it has a natural positive direction — the one for which n points away from the interior, i.e., points toward the outside. We shall always understand that the closed surface has been oriented this way, unless otherwise specified.

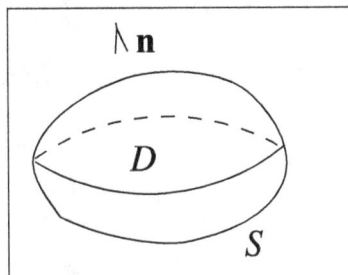

We now generalize to 3-space the normal form of Green's theorem.

Let F(x, y, z) = M i + N j + P k be a vector field differentiable in some region D. By the divergence of F we mean the scalar function div F of three variables defined in D by:

$$div\,F = \frac{\partial M}{\partial x} + \frac{\partial N}{\partial y} + \frac{\partial P}{\partial z}.$$

The divergence theorem: Let S be a positively-oriented closed surface with interior D, and let F be a vector field continuously differentiable in a domain contatining D. Then:

$$\iint_S F \cdot dS = \iiint_D div\,F\,dV$$

We write dV on the right side, rather than dx dy dz since the triple integral is often calculated in other coordinate systems, particularly spherical coordinates. The theorem is sometimes called Gauss' theorem.

Physically, the divergence theorem is interpreted just like the normal form for Green's theorem. Think of F as a three-dimensional flow field. The surface integral represents the mass transport rate across the closed surface S, with flow out of S considered as positive, flow into S as negative.

Proof of the Divergence Theorem

Let $\widehat{F}$ be a smooth vector field defined on a solid region V with boundary surface A oriented outward. We wish to show that:

$$\int_A \vec{F} \cdot d\vec{A} = \int_V div\,\vec{F}\,dV.$$

For the Divergence Theorem, we use the same approach as we used for Green's Theorem; first prove the theorem for rectangular regions, then use the change of variables formula to prove it for regions parameterized by rectangular regions, and finally paste such regions together to form general regions.

Proof for Rectangular Solids with Sides Parallel to the Axes

Consider a smooth vector field $\widehat{F}$ defined on the rectangular solid $V: a \le x \le b, c \le y \le d, e \le z \le f$.

We start by computing the flux of $\widehat{F}$ through the two faces of V perpendicular to the x-axis, A_1 and A_2, both oriented outward:

$$\int_{A_1} \vec{F} \cdot d\vec{A} + \int_{A2} \vec{F} \cdot d\vec{A} = -\int_e^f \int_c^d F_1(a, y, z)\,dy\,dz + \int_c^f \int_c^d F_1(b, y, z)\,dy\,dz$$

$$= \int_c^f \int_c^d \left(F_1(b, y, z) - F_1(a, y, z) \right) dy\,dz.$$

By the Fundamental Theorem of Calculus:

$$F_1(b, y, z) - F_1(a, y, z) = \int_a^b \frac{\partial F_1}{\partial x}\,dx,$$

So,

$$\int_{A_1} \vec{F} \cdot d\vec{A} + \int_{A_2} \vec{F} \cdot d\vec{A} = \int_e^f \int_c^d \int_a^b \frac{\partial F_1}{\partial x} \, dx \, dy \, dz = \int_V \frac{\partial F_1}{\partial x} \, dV.$$

By a similar argument, we can show:

$$\int_{A_3} \vec{F} \cdot d\vec{A} + \int_{A_4} \vec{F} \cdot d\vec{A} = \int_V \frac{\partial F_2}{\partial y} \, dV \text{ and } \int_{A_5} \vec{F} \cdot d\vec{A} + \int_{A_6} \vec{F} \cdot d\vec{A} = \int_V \frac{\partial F_3}{\partial z} \, dV.$$

Adding these we get:

$$\int_A \vec{F} \cdot d\vec{A} = \int_V \left(\frac{\partial F_1}{\partial x} + \frac{\partial F_2}{\partial y} + \frac{\partial F_3}{\partial z} \right) dV = \int_V \operatorname{div} \vec{F} \, dV.$$

This is the Divergence Theorem for the region V.

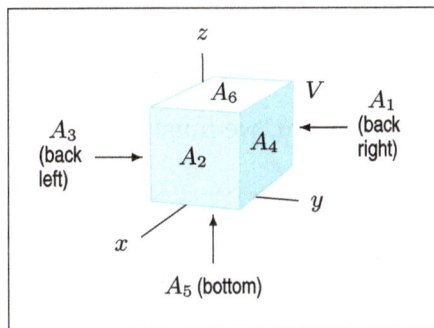

Rectangular solid V in xyz-space.

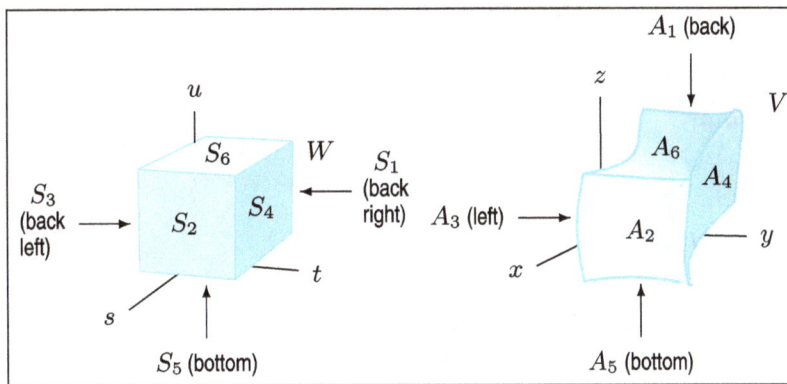

A rectangular solid W in stu-space and the corresponding curved solid V in xyz-space.

Proof for Regions Parameterized by Rectangular Solids

Now suppose we have a smooth change of coordinates:

$x = x(s, t, u), y = y(s, t, u), z = z(s, t, u).$

Consider a curved solid V in xyz-space corresponding to a rectangular solid W in stu-space. We suppose that the change of coordinates is one-to-one on the interior of W, and that its Jacobian determinant is positive on W. We prove the Divergence Theorem for V using the Divergence Theorem for W.

Let A be the boundary of V. To prove the Divergence Theorem for V, we must show that:

$$\int_A \vec{F} \cdot d\vec{A} = \int_V div \vec{F} \, dV.$$

First we express the flux through A as a flux integral in stu-space over S, the boundary of the rectangular region W. In vector notation the change of coordinates is:

$$\vec{r} = \vec{r}(s,t,u) = x(s,t,u)\vec{i} + y(s,t,u)\vec{j} + z(s,t,u)\vec{k}.$$

The face A_1 of V is parameterized by,

$$\vec{r} = \vec{r}(a,t,u), \qquad c \leq t \leq d, \, e \leq u \leq f,$$

so on this face,

$$d\vec{A} = \pm \frac{\partial \vec{r}}{\partial t} \times \frac{\partial \vec{r}}{\partial u} \, dt \, du.$$

In fact, in order to make $d\vec{A}$ point outward, we must choose the negative sign. Thus, if S_1 is the face $s = a$ of W

$$\int_{A_1} \vec{F} \cdot d\vec{A} = -\int_{S_1} \vec{F} \cdot \frac{\partial \vec{r}}{\partial t} \times \frac{\partial \vec{r}}{\partial u} \, dt \, du.$$

The outward pointing area element on S_1 is $\vec{S} = -\vec{i} \, dt \, du$. Therefore, if we choose a vector field $\vec{G}$ on stu-space whose component in the s-direction is,

$$G_1 = \vec{F} \cdot \frac{\partial \vec{r}}{\partial t} \times \frac{\partial \vec{r}}{\partial u},$$

we have,

$$\int_{A_1} \vec{F} \cdot d\vec{A} = \int_{S_1} \vec{G} \cdot d\vec{S}.$$

Similarly, if we define the t and u components of $\vec{G}$ by,

$$G_2 = \vec{F} \cdot \frac{\partial \vec{r}}{\partial u} \times \frac{\partial \vec{r}}{\partial s} \quad and \quad G_3 = \vec{F} \cdot \frac{\partial \vec{r}}{\partial s} \times \frac{\partial \vec{r}}{\partial t},$$

then,

$$\int_{A_i} \vec{F} \cdot d\vec{A} = \int_{S_i} \vec{G} \cdot d\vec{S}, i = 2, ..., 6.$$

Adding the integrals for all the faces, we find that:

$$\int_A \vec{F} \cdot d\vec{A} = \int_S \vec{G} \cdot d\vec{S}.$$

Since we have already proved the Divergence Theorem for the rectangular region W we have,

$$\int_S \vec{G} \cdot d\vec{S} = \int_W div\ \vec{G}\ dW,$$

where,

$$div\ \vec{G} = \frac{\partial G_1}{\partial s} + \frac{\partial G_2}{\partial t} + \frac{\partial G_3}{\partial u}.$$

$$\frac{\partial G_1}{\partial s} + \frac{\partial G_2}{\partial t} + \frac{\partial G_3}{\partial u} = \left|\frac{\partial(x,y,z)}{\partial(s,t,u)}\right|\left(\frac{\partial F_1}{\partial x} + \frac{\partial F_2}{\partial y} + \frac{\partial F_3}{\partial x}\right).$$

So, by the three-variable change of variables formula,

$$\int_V div\ \vec{F}\ dV = \int_V \left(\frac{\partial F_1}{\partial x} + \frac{\partial F_2}{\partial y} + \frac{\partial F_3}{\partial z}\right) dx\ dy\ dz$$

$$= \int_W \left(\frac{\partial F_1}{\partial x} + \frac{\partial F_2}{\partial y} + \frac{\partial F_3}{\partial z}\right)\left|\frac{\partial(x,y,z)}{\partial(s,t,u)}\right| ds\ dt\ du$$

$$= \int_W \left(\frac{\partial F_1}{\partial s} + \frac{\partial G_2}{\partial t} + \frac{\partial G_3}{\partial u}\right) ds\ dt\ du$$

$$= \int_W div\ \vec{G}\ dW.$$

we have shown that,

$$\int_A \vec{F} \cdot d\vec{A} = \int_S \vec{G} \cdot d\vec{S}$$

and

$$\int_V div\ \vec{F}\ dV = \int_V div\ \vec{G}\ dW.$$

By the Divergence Theorem for rectangular solids, the right-hand sides of these equations are equal, so the left-hand sides are equal also. This proves the Divergence Theorem for the curved region V.

Pasting Regions Together

As in the proof of Green's Theorem, we prove the Divergence Theorem for more general regions by pasting smaller regions together along common faces. Suppose the solid region V is formed by pasting together solids V_1 and V_2 along a common face, as in figure.

The surface A which bounds V is formed by joining the surfaces A_1 and A_2 which bound V_1 and V_2, and then deleting the common face. The outward flux integral of a vector field $\vec{F}$ through A_1 includes the integral across the common face, and the outward flux integral of $\vec{F}$ through A_2 includes the integral over the same face, but oriented in the opposite direction.

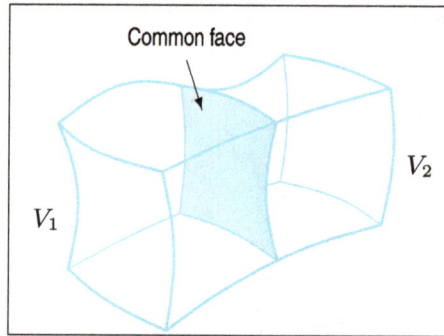

Region V formed by pasting together V_1 and V_2.

Thus, when we add the integrals together, the contributions from the common face cancel, and we get the flux integral through A. Thus we have:

$$\int_A \vec{F} \cdot d\vec{A} = \int_{A_1} \vec{F} \cdot d\vec{A} + \int_{A_2} \vec{F} \cdot d\vec{A}.$$

But we also have,

$$\int_V div\,\vec{F}\ dV = \int_V div\,\vec{F}dv + \int_{V_2} div\,\vec{F}\ dV.$$

So the Divergence Theorem for V follows from the Divergence Theorem for V_1 and V_2. Hence we have proved the Divergence Theorem for any region formed by pasting together regions that can be smoothly parameterized by rectangular solids.

Example: Let V be a spherical ball of radius 2, centered at the origin, with a concentric ball of radius 1 removed. Using spherical coordinates, show that the proof of the Divergence Theorem we have given applies to V.

Solution: We cut V into two hollowed hemispheres like the one shown in figure, W. In spherical coordinates, W is the rectangle $1 \le \rho \le 2$, $0 \le \varphi \le \pi$, $0 \le \theta \le \pi$. Each face of this rectangle becomes part of the boundary of W. The faces $\rho = 1$ and $\rho = 2$ become the inner and outer hemispherical surfaces that form part of the boundary of W. The faces $\theta = 0$ and $\theta = \pi$ become the two halves of the flat part of the boundary of W. The faces $\varphi = 0$ and $\varphi = \pi$ become line segments along the z-axis. We can form V by pasting together two solid regions like W along the flat surfaces where $\theta = $ constant.

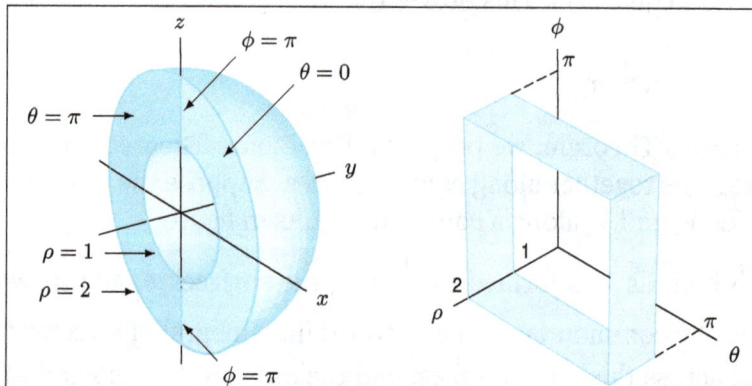

The hollow hemisphere W and the corresponding rectangular region in $\rho\theta\varphi$ – space.

Gradient Theorem

In one-variable calculus, the fundamental theorem of calculus was a useful tool for evaluating integrals. If you are integrating a function g(t) and it turns out that the function is the derivative of another function g(t) = G'(t), then integrating the function g(t) is simple. The integral of g is just the difference in the values of G(t) at the endpoints. We could write the result as:

$$\int_a^b G'(t)\,dt = G(b) - G(a).$$

For line integrals of vector fields, there is a similar fundamental theorem. In some cases, we can reduce the line integral of a vector field F along a curve C to the difference in the values of another function f evaluated at the endpoints of C,

$$\int_C F \cdot ds\, f(Q) - f(P),$$

where C starts at the point P and ends at the point Q. If we let $F : R^n \to R^n$ be a n-dimensional vector field, then it must be that $f : R^n \to R$ is a scalar-valued function, as the line integral evaluates to a single number.

This sounds good, but there is an important catch: it will only work for integrating specials kinds of vector fields. Clearly, equation $\int_C F \cdot ds\, f(Q) - f(P)$, could possibly be true only if the line integral along C depends only on the endpoints of C and doesn't depend on the particular path the C takes. In other words, we can hope for a similar fundamental theorem for line integrals only if the vector field is conservative (also called path-independent). If the vector field F is path-dependent, then it will be impossible to reduce its line integral to values of a function at the path endpoints.

We can easily derive what a conservative vector field should look like and in the process obtain our fundamental theorem for line integrals. Let the curve C from point P to Q be parametrized by c(t) for a < t < b. This means P = c(a), Q = c(b), and the line integral $\int_C F \cdot ds$ can be written as $\int_a^b F(c(t)) \cdot c'(t)\,dt$. The desired relaationship between F and f described by equation $\int_C F \cdot ds\, f(Q) - f(P)$, can be rewritten as:

$$\int_a^b F(c(t)) \cdot c'(t)\,dt = f(c(b)) - f(c(a)).$$

In this form, equation $\int_a^b F(c(t)) \cdot c'(t)\,dt = f(c(b)) - f(c(a))$ is starting to look like our original equation $\int_a^b G'(t)\,dt = G(b) - G(a)$ for the fundamental theorem of calculus. If we let G(t) = f(c(t)), then the right hand side of equation $\int_a^b F(c(t)) \cdot c'(t)\,dt = f(c(b)) - f(c(a))$ is indeed

G(b) – G(a). For equation $\int_a^b F\big(c(t)\big)\cdot c'(t)\,dt = f\big(c(b)\big) - f\big(c(a)\big)$ to be valid, we just need, F(c(t)) · c'(t) to be equal to G'(t).

Since G(t) = f(c(t)) is a composition of functions, we can compute its derivative using the chain rule. For $c : \mathbb{R} \to \mathbb{R}^n$ and $f : \mathbb{R}^n \to \mathbb{R}$, the chain rule can be written as:

G'(t) = Df(c(t))Dc(t) = ∇f(c(t)) · c'(t).

The gradient vector ∇f is just the vector form of the 1×n derivative matrix Df, and the derivative of a parametrized curve is the tangent vector c'(t).

The expression for G'(t) is in exactly the form we need. G'(t) will be equal to F(c(t)) · c'(t) under the condition that we find a function f so that the vector field F is the gradient ∇f. Let's assume that F=∇f. Then, finally, by the one-variable fundamental theorem of calculus of equation $\int_a^b G'(t)\,dt = G(b) - G(a)$. We know that the desired relationship of equation

$\int_a^b F\big(c(t)\big)\cdot c'(t)\,dt = f\big(c(b)\big) - f\big(c(a)\big)$ is valid in the form:

$$\int_a^b \nabla f\big(c(t)\big)\cdot c'(t)\,dt = f\big(c(b)\big) - f\big(c(a)\big).$$

Rewriting this expression in terms of the original curve C from point P to point Q, we obtain the gradient theorem for line integrals:

$$\int_C \nabla f \cdot ds = f(Q) - f(P).$$

This theorem is also called the fundamental theorem for line integrals, as it is a generalization of the one variable fundamental theorem of calculus of equation $\int_a^b G'(t)\,dt = G(b) - G(a)$. to line integrals along a curve.

Using the Gradient Theorem

The gradient theorem makes evaluating line integrals $\int_C F \cdot ds$ very simple, if we happen to know that F = ∇f. The function f is called the potential function of F. Typically, though you just have the vector field F, and the trick is to know if a potential function exists and, if so, how find it.

It is clear from the above function that a vector field has a potential function only if it is conservative (or path-independent). It turns out the converse is true as well, so that a potential function f exists satisfying ∇f=F if and only if F is conservative. So, the two steps for using the gradient theorem to evaluate a line integral $\int_C F \cdot ds$ are,

- Determine if F is conservative, and

- Find the potential function f if F is conservative.

With the potential function f in hand, evaluating $\int_C F \cdot ds$ is as simple as calculating the values of f at the endpoints of C and subtracting, according to the gradient theorem of equation

$$\int_C \nabla f \cdot ds = f(Q) - f(P).$$

Example of using the Gradient Theorem

If a vector field F is a gradient field, meaning F = ∇f for some scalar-valued function f, then we can compute the line integral of F along a curve C from some point a to some other point b as:

$$\int_C F \cdot ds = f(b) - f(a).$$

This integral does not depend on the entire curve C; it depends on only the endpoints a and b. If we replaced C by another curve with the same endpoints, the integral would be unchanged. Hence F is conservative (also called path-independent).

As an example, consider f(x, y) = xy², and let F(x, y) = ∇f (x, y) = (y², 2xy). Since we wrote F as a gradient, we know that F must be conservative.

What is $\int_C F \cdot ds$ where C is the path c(t) = (t², 2(t − 2)³) for 1 ≤ t ≤ 3? The starting point is a = c(1) = (1,−2), and the ending pont is b = c(3) = (9, 2). Hence the integral must be:

$$\int_C F \cdot ds = f(9,2) - f(1,-2)$$
$$= 9(2^2) - 1(-2)^2 = 36 - 4 = 32.$$

We could also compute $\int_C F \cdot ds$ the direct way using the parametrization c(t). The integral isn't difficult as it is just a polynomial, but it is messy:

$$\int_C F \cdot ds = \int_a^b F(c(t)) \cdot c'(t) dt$$
$$= \int_1^3 F(t^2, 2(t-2)^3) \cdot (2t, 6(t-2)^2) dt$$
$$= \int_1^3 \left((2(t-2)^3)^2, 2t^2 2(t-2)^3 \right) \cdot (2t, 6(t-2)^2) dt$$
$$= \int_1^3 8(t(t-2)^6 + 3t^2 (t-2)^5) dt$$
$$= 8(32t^2 - 96t^3 + 120t^4 - 80t^5 + 30t^6 - 6t^7 + t^2/2)\Big|_1^3$$
$$= 36 - 4 = 32,$$

which agrees with the first answer.

We should get the same answer for any path from (1, −2) to (9, 2). Since (9, 2) − (1, −2) = (8, 4), we let the curve B be the straight line path parametrized by p(t) = (1, −2) + t(8, 4) = (1 + 8t, 4t − 2) for $0 \le t \le 1$. This change is not a reparametrization of C. The curve C was not a straight line, so B is a completely different curve, as you can see in the below figure, where C is shown in blue and B is shown in green.

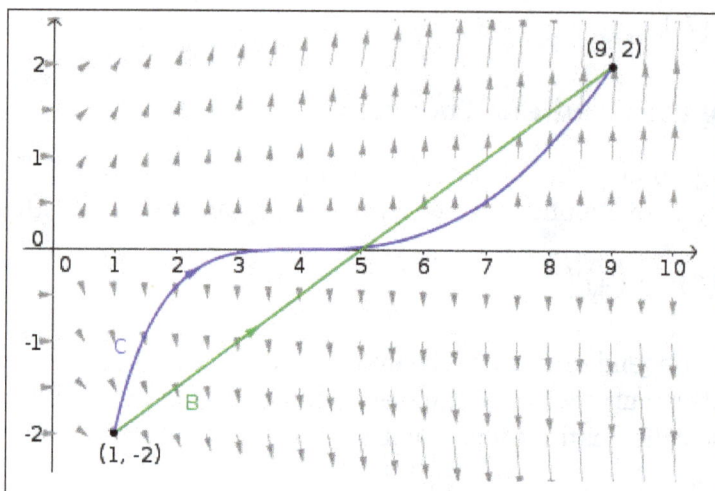

The integral along B is,

$$\int_B F \cdot ds = \int_0^1 F\big(p(t)\big) \cdot p'(t)\, dt$$

$$= \int_0^1 F(1+8t, 4t-2) \cdot (8,4)\, dt$$

$$= \int_0^1 \Big((4t-2)^2, 2(4t-2)(1+8t)\Big) \cdot (8,\ 4)\, dt$$

$$= \int_0^1 \Big(4-16t+16t^2, -4-24t+64t^2\Big) \cdot (8,4)\, dt$$

$$= \int_0^1 16\Big(1-14t+24t^2\Big)\, dt$$

$$= 16\Big(t-7t^2+8t^3\Big)\Big|_0^1 = 32.$$

Indeed, we got the same answer again.

Stokes' Theorem

Stokes' Theorem states that the line integral of a closed path is equal to the surface integral of any capping surface for that path, provided that the surface normal vectors point in the same general direction as the right-hand direction for the contour:

$$\oint_C \vec{F} \cdot d\vec{r} = \iint_S \big(\nabla \times \vec{F}\big) \cdot d\vec{S},$$

Intuitively, imagine a "capping surface" that is nearly flat with the contour. The curl is the microscopic circulation of the function on tiny loops within that surface, and their sum or integral results in canceling out all the internal circulation paths, leaving only the integration over the outer-most path. This remains true no matter how the capping surface is expanded, provided that the contour remains as its boundary.

Sometimes the circulation (the left side above) is easier to compute; other times the express-es the surface integral of the curl of vector field is easier to computer (particularly when it is zero).

Stated another way, Stokes' Theorem equates the line integral of a vector fields to a surface inte-gral of the same vector field. For this identity to be true, the direction of the vector normal n must obey the right-hand rule for the direction of the contour, i.e., when walking along the contour the surface must be on your left.

This is an extension of Green's Theorem to surface integrals, and is also the analog in two dimen-sions of the Divergence Theorem. The above formulation is also called as the "Curl Theorem," to distinguish it from the more general form of the Stokes' Theorem described below.

Stokes' Theorem is useful in calculating circulation in mechanical engineering. A conservative field has a circulation (line integral on a simple, closed curve) of zero, and application of the Stokes' Theorem to such a field proves that the curl of a conservative field over the enclosed surface must also be zero.

General Form

In its most general form, this theorem is the fundamental theorem of Exterior Calculus, and is a generalization of the Fundamental Theorem of Calculus. It states that if M is an oriented piecewise smooth manifold of dimension k and ω is a smooth $(k - 1)$-form with compact support on M, and ∂M denotes the boundary of M with its induced orientation then,

$$\int_m d\omega = \oint_{\partial M} \omega$$

where d is the exterior derivative.

There are a number of well-known special cases of Stokes' theorem, including one that is re-ferred to simply as "Stokes' theorem" in less advanced treatments of mathematics, physics, and engineering:

• When k = 1, and the terms appearing in the theorem are translated into their simpler form, this is just the Fundamental Theorem of Calculus.

• When k = 3, this is often called Gauss' Theorem or the Divergence Theorem and is useful in vector calculus:

$$\iiint_R \left(\nabla \cdot \vec{\omega} \right) dV = \iint_S \vec{\omega} \cdot d\vec{A}$$

Where R is some region of 3-space, S is the boundary surface of R, the triple integral denotes vol-ume integration over R with dV as the volume element, and the double integral denotes surface integration over S with $d\vec{A}$ as the oriented normal of the surface element. The ∇ on the left side is the divergence operator, and the $\cdot$ on the right side is the vector dot product.

- When k=2, this is often just called Stokes' Theorem:

$$\iint_S \left(\nabla \cdot \vec{\omega}\right) \cdot d\vec{A} = \oint_E \vec{\omega} \cdot \vec{dl}$$

Here S is a surface, E is the boundary path of S, and the single integral denotes path integration around E with $\vec{dl}$ as the length element. The $\nabla \times$ on the left side is the curl operator.

These last two examples (and Stokes' theorem in general) are the subject of vector calculus. They play important roles in electrodynamics. The divergence and curl operations are cornerstones of Maxwell's equations.

Stokes' Theorem is a lower-dimension version of the Divergence Theorem, and a higher-dimension version of Green's Theorem.

Example: Let C be the closed curve illustrated below.

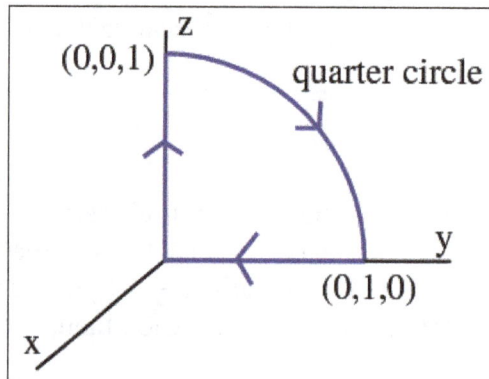

For F(x, y, z) = (y, z, x), compute,

$$\int_C F \cdot ds$$

using Stokes' Theorem.

Solution: Since we are given a line integral and told to use Stokes' theorem, we need to compute a surface integral,

$$\iint_C curl\, F \cdot ds$$

where S is a surface with boundary C. We have freedom to choose any surface S, as long as we orient it so that C is a positively oriented boundary.

In this case, the simplest choice for S is clear. Let S be the quarter disk in the yz-plane.

Given the orientation of the curve C, we need to choose the surface normal vector n to point in which direction? By the right hand rule criterion, the normal vector should point toward the negative side of the x-axis.

We need to calculate the curl of F. We can calculate the curl as using the notation:

$$curl(F) = \nabla \times F = \nabla \times (y, z, x)$$

$$= \begin{vmatrix} i & j & k \\ \dfrac{\partial}{\partial x} & \dfrac{\partial}{\partial y} & \dfrac{\partial}{\partial z} \\ y & z & x \end{vmatrix}$$

$$= i\left(\frac{\partial}{\partial y}x - \frac{\partial}{\partial z}z\right) - j\left(\frac{\partial}{\partial x}x - \frac{\partial}{\partial z}y\right)$$

$$+ k\left(\frac{\partial}{\partial x}z - \frac{\partial}{\partial y}y\right)$$

$$= i(-1) - j(1) + k(-1)$$

$$= (-1, -1, -1)$$

Next, parameterize the surface (the quarter disk) by:

$$\Phi(r, \theta) = (0, r\cos\theta, r\sin\theta)$$

for $0 \leq r \leq 1$ and $0 \leq \theta \leq \pi/2$.

Calculate the normal vector (we don't need to normalize it to the unit normal vector n):

$$\frac{\partial \Phi}{\partial r} = (0, \cos\theta, \sin\theta)$$

$$\frac{\partial \Phi}{\partial \theta} = (0, -r\sin\theta, r\cos\theta)$$

$$\frac{\partial \Phi}{\partial r} \times \frac{\partial \Phi}{\partial \theta} = i\left(r\cos^2\theta + r\sin^2\theta\right) = ri$$

Is the surface oriented properly? The normal vector points in the positive x-direction. But we need it to point it negative x-direction. Therefore, the surface is not oriented properly if we were to choose this normal vector.

To orient the surface properly, we must instead use the normal vector:

$$\frac{\partial \Phi}{\partial \theta} = \frac{\partial \Phi}{\partial r} = -ri.$$

At this point, we can already see that the integral $\iint_S F \cdot dS$ should be positive. The vector field curl F = (−1, −1, −1) and the normal vector (−r, 0, 0) are pointing in a similar direction.

Now, we have all pieces together to compute the integral:

$$\int_C F \cdot ds = \iint_S curl\ F \cdot dS$$

$$= \int_0^1 \int_0^{\pi/2} curl\ F\left(\Phi(r,\theta)\right) \cdot \left(\frac{\partial \Phi}{\partial \theta}(r,\theta) \times \frac{\partial \Phi}{\partial r}(r,\theta)\right) d\theta\, dr$$

$$= \int_0^1 \int_0^{\pi/2} (-1,-1,-1) \cdot (-r,0,0)\, d\theta\, dr$$

$$= \int_0^1 \int_0^{\pi/2} r\, d\theta\, dr = \frac{\pi}{4}$$

Example: Just for verification, we can compute the line $\int_C F \cdot dS$ directly.

We need to parametrize C. We'll do it by dividing *C* into three parts.

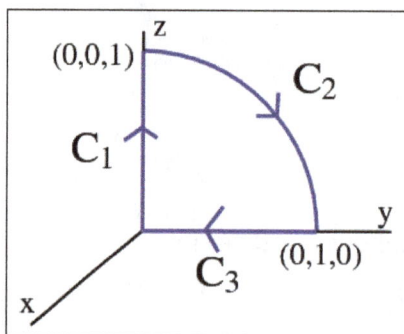

We'll use the fact that,

$$\int_C F \cdot dS = \int_{C_1} F \cdot dS + \int_{C_2} F \cdot dS + \int_{C_3} F \cdot dS$$

Recall F(x, y, z) = (y, z, x)

First we'll compute the integral over C_1. Parameterize it by:

c(t) = (0, 0, t), 0 ≤ t ≤ 1.

Since c′(t) = (0, 0, 1), we compute that,

$$F(c(t)) \cdot c'(t) = F(0,0,t) \cdot (0,0,1)$$
$$= (0,t,0) \cdot (0,0,1)$$
$$= 0$$

Therefore:

$$\int_C F \cdot dS = \int_0^1 F(c(t)) \cdot c'(t) dt = 0.$$

The integral for C_3 is similar.

$$\int_{C_3} F \cdot dS = 0$$

Last, we'll compute the integral over C_2. Parameterize C_2 as,

$$c(t) = (0, \sin t, \cos t), \quad 0 \le t \le \pi/2,$$

so that $c'(t) = (0, \cos t, -\sin t)$. We then compute:

$$\int_{C_3} F \cdot dS = \int_0^{\pi/2} F\big(c(t)\big) \cdot c'(t)\, dt$$

$$= \int_2^{\pi/2} F(0, \sin t, \cos t) \cdot (0, \cos t, -\sin t)\, dt$$

$$= \int_2^{\pi/2} (\sin t, \cos t, 0) \cdot (0, \cos t, -\sin t)\, dt$$

$$= \int_2^{\pi/2} \cos^2 t\, dt$$

$$= \int_2^{\pi/2} \frac{1 + \cos 2t}{2}\, dt$$

$$= \frac{t}{2} + \frac{\sin 2t}{4}\Big|_0^{\pi/2} = \frac{\pi}{4}.$$

Therefore,

$$\int_{C_3} F \cdot dS = \frac{\pi}{4}$$

in agreement with our Stokes' theorem answer.

Example: We often present Stoke's theorem problems as we did above. We give a curve C and expect you to compute the surface integral over some surface S with boundary C. In general, one can pick many surfaces. But, sometimes, there is a surface that is "obviously" the best one.

One special case where this is relatively easy is when C lies in a plane. This is especially easy when that plane is parallel to a coordinate plane, as in the following example.

Let's say you want to use Stokes' theorem to compute $\int_{C_3} F \cdot dS$ where C is polygon path connecting the following points: $(1, 1, 0)$, $(3, 1, 4)$, $(1, 1, 5)$, $(-1, 1, 1)$.

Does this curve lie in a plane? Yes, in the plane $y = 1$. The figure below is just the plane $y = 1$.

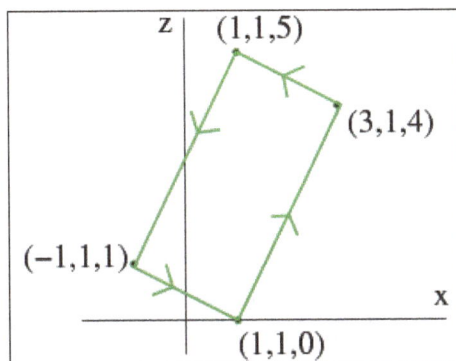

If one coordinate is constant, then curve is parallel to a coordinate plane. (The xz-plane for above example). For Stokes' theorem, use the surface in that plane. For our example, the natural choice for S is the surface whose x and z components are inside the above rectangle and whose y component is 1.

Example: In other cases, a surface is given explicitly in the problem.

Compute $\int_{C_3} F \cdot dS$, where C is the curve in which the cone $z^2 = x^2 + y^2$ intersects the plane $z = 1$.

(Oriented counter clockwise viewed from positive z-axis).

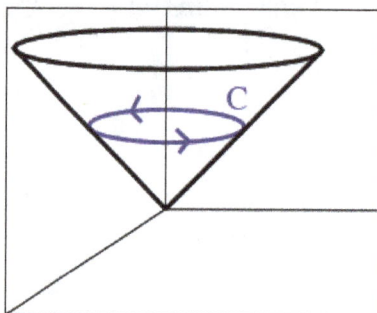

$$\int_{C_3} F \cdot dS = \iint_S curl\, F \cdot dS$$

for what surface S?

In this case, there are two natural choices for the surface. You could use the portion of the plane or the portion of the cone illustrated below.

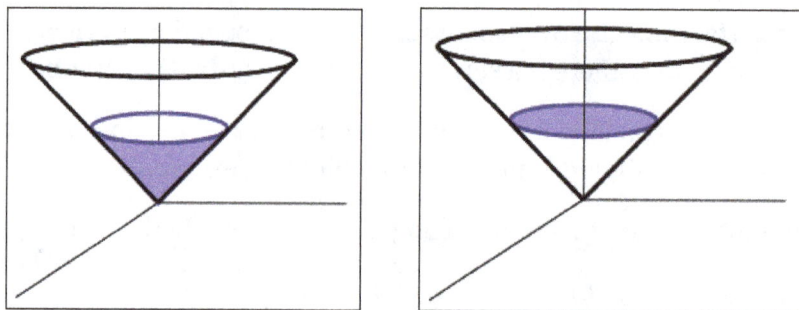

Let P be the portion of the plane $z = 1$ with $x^2 + y^2 < 1$ with upward pointing normal. Let Q be the portion of the cone $z^2 = x^2 + y^2$ with $0 < z < 1$ with upward angling normal.

How do $\iint_P curl\, F \cdot dS$ and $\iint_Q curl\, F \cdot dS$ compare? They are the same. For both surfaces, C is a positive oriented boundary.

Continue example: Let,

$$F(x, y, z) = \left(\sin x - \frac{y^3}{3}, \cos y + \frac{x^3}{3}, xyz \right)$$

Compute $\int_C F \cdot dS$.

One can show that curl (F) = $(xz, -yz, x^2 + y^2)$.

Use surface P, parameterized by,

$$\Phi(r,\theta) = (r\cos\theta, \ r\sin\theta, \ 1)$$

for $0 \le r \le 1, 0 \le \theta \le 2\pi$. Then normal vector is,

$$\frac{\partial \Phi}{\partial r} \times \frac{\partial \Phi}{\partial \theta} = (0,0,r),$$

which points in the correct direction, as mentioned above.

$$\iint_P curl \ F \cdot dS = \int_0^1 \int_0^{2\pi} curl \left(F\left(r\cos\theta, r\sin\theta, 1\right)\right) \cdot (0,0,r) \, d\theta \, dr$$
$$= \int_0^1 \int_0^{2\pi} \left(r\cos\theta, -r\sin\theta, r^2 \right) \cdot (0,0,r) \, d\theta \, dr$$
$$= \int_0^1 \int_0^{2\pi} r^3 \, d\theta \, dr$$
$$= \int_0^1 2\pi r^3 \, dr = \frac{\pi}{2}$$

Proper Orientation for Stokes' Theorem

One important subtlety of Stokes' theorem is orientation. We need to be careful about orientating the surface (which is specified by the normal vector n) properly with respect to the orientation of the boundary (which is specified by the tangent vector). Remember, changing the orientation of the surface changes the sign of the surface integral. If we choose the wrong n (i.e., the wrong orientation), we could be off by a minus sign.

Look at the below image from the Stokes' theorem introduction, where the "microscopic circulation" is sketched by green circles on the surface. Notice how the arrows on the little green circles (indicating the "microscopic circulation") are aligned with the red arrow indicating the direction of the curve C. If, for example, the arrows on the green circles were going the other direction, the green circles and the red curve wouldn't match, and we'd be off by a minus sign.

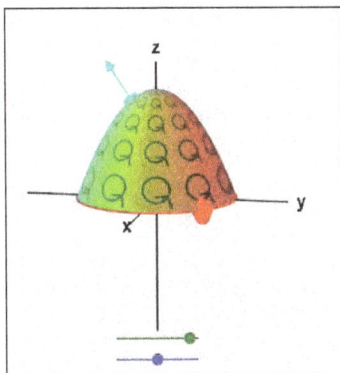

Macroscopic and microscopic circulation in three dimensions: The relationship between the macroscopic circulation of a vector field F around a curve (red boundary of surface) and the microscopic circulation of F (illustrated by small green circles) along a surface in three dimensions must hold for any surface whose boundary is the curve. No matter which surface you choose (change by dragging the green point on the top slider), the total microscopic circulation of F along the surface must equal the circulation of F around the curve. (We assume that the vector field F is defined everywhere on the surface.) You can change the curve to a more complicated shape by dragging the blue point on the bottom slider, and the relationship between the macroscopic and total microscopic circulation still holds. The surface is oriented by the shown normal vector (moveable cyan arrow on surface), and the curve is oriented by the red arrow.

Looking from the positive z-axis, both the green circles and the red curve indicate counter clockwise circulation. To define the orientation for Green's theorem, this was sufficient. We simply insisted that you orient the curve C in the counter clockwise fashion. For Stokes' theorem, we cannot just say "counter clockwise," since the orientation that is counter clockwise depends on the direction from which you are looking. If you take the image and rotate it 180° so that you are looking at it from the negative z-axis, the same curve would look like it was oriented in the clockwise fashion. Since the green circles would also look like they are oriented in a clockwise fashion, you can still see that the green circles and the red curve match.

Remember, too, that the curve C can be floating or twisted in any direction. It doesn't have to look as simple as in the above examples. Thankfully, choosing the correct orientation doesn't have to be too difficult if you remember the right hand rule. If you look at your right hand from the side of your thumb, your fingers curl in the counter clockwise direction. Think of your thumb as the normal vector n of a surface. If your thumb points to the positive side of the surface, your fingers indicate the circulation corresponding to curl F· n. In the image, the normal vector corresponding to the orientation of the green circles is shown as a cyan arrow. If you place the thumb of your right hand so that it points in the direction of the cyan normal vector, the fingers of your right hand curl in the direction corresponding to the orientation of the green circles.

With your thumb oriented corresponding to the cyan normal vector, move your hand along the surface toward its edges. When your fingers are next to the boundary of the surface, the red curve C must be oriented (by red arrow) to go around the same direction your fingers are pointing. If the relationship between the normal vector n and the orientation of C doesn't match the relationship between the thumb and fingers of your right hand, you'll be off by a minus sign when trying to apply Stokes' theorem.

Another way of thinking about the proper orientation is the following. Imagine that you are walking on the positive side of the surface (i.e, the side with the cyan normal vector in the image). If you walk near the edge of the surface in the direction corresponding to the orientation of C, then surface must be to your left and the edge C must be to your right.

When the curve C and the surface S are oriented as described above so that Stokes' theorem applies, we say that C is a positively oriented boundary of S.

Green's Theorem

Green's theorem states that a line integral around the boundary of a plane region D can be computed as a double integral over D. More precisely, if D is a "nice" region in the plane and C is the boundary of D with C oriented so that D is always on the left-hand side as one goes around C (this is the positive orientation of C) then:

$$\int_C P\,dx + Q\,dy = \iint_D \left(\frac{\partial Q}{\partial x} - \frac{\partial P}{\partial y} \right) dx\,dy$$

If the partial derivatives of P and Q are continuous on D.

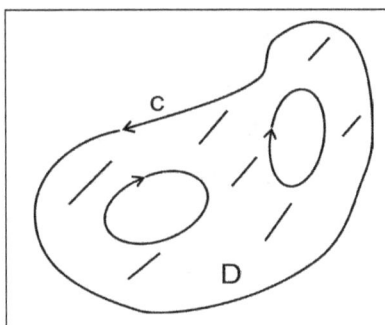

Regions that are simultaneously of type I and II are "nice" regions, i.e., Green's theorem is true for such regions. The next two propositions prove this.

Theorem: If D is a region of type I then:

$$\int_C P\,dx = \iint_D \frac{\partial P}{\partial y}\,dx\,dy.$$

Proof: If $D = \{(x, y) \mid a \le x \le b, f(x) \le y \le g(x)\}$ with $f(x)$, $g(x)$ continuous on $a \le x \le b$, we have $C = C_1 + C_2 + C_3 + C_4$, where C_1, C_2, C_3, C_4 are as shown below.

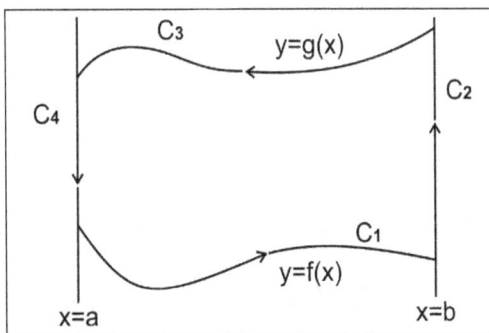

Since the region is of type I, we have:

$$\iint_D \frac{\partial P}{\partial y}\,dx\,dy = \int_a^b \left[P(x, g(x)) - P(x, f(x)) \right] dx.$$

Using the standard parametrizations of C_1 and C_3, we have:

$$\int_a^b P(x, f(x))\,dx = \int_{C_1} P\,dx, \quad \int_a^b P(x, g(x))\,dx = -\int_{C_3} P\,dx$$

We thus obtain,

$$\iint_D \frac{\partial P}{\partial y}\,dx\,dy = \int_{C_1} P\,dx, \quad \int_{C_3} P\,dx = \int_C P\,dx$$

since the line integral of $P\,dx$ is zero on C_2 and C_4 as x is constant there.

Theorem: If D is a region of type II then:

$$\int_C Q\,dy = \iint_D \frac{\partial Q}{\partial x}\,dx\,dy.$$

Proof: If $D = \{(x, y) \mid h(y) \le x \le k(y), c \le y \le d\}$ with $h(y)$, $k(y)$ continuous on $c \le y \le d$, we have $C = C_1 + C_2 + C_3 + C_4$, where C_1, C_2, C_3, C_4 are as shown below.

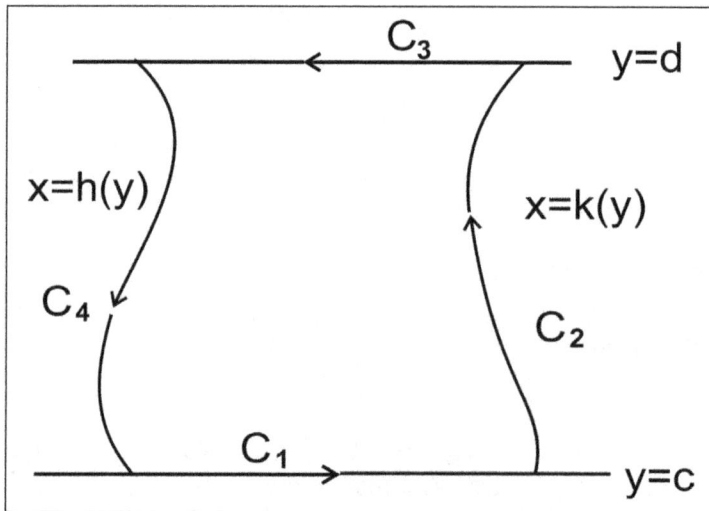

Since the region is of type II, we have:

$$\iint_D \frac{\partial Q}{\partial x}\,dx\,dy = \int_c^d \left[Q(k(y), y) - Q(h(y), y) \right]dy.$$

Using the standard parametrizations of C_2 and C_4, we have:

$$\int_c^d Q(k(y), y)\,dy = \int_{C_2} Q\,dy, \quad \int_c^d Q(h(y), y)\,dx = -\int_{C_4} Q\,dy$$

We thus obtain,

$$\iint_D \frac{\partial Q}{\partial x}\,dx\,dy = \int_{C_2} Q\,dy + \int_{C_4} Q\,dy = \int_C Q\,dy$$

since the line integral of Q dx is zero on C_1 and C_3 as y is constant there. Putting these two theorems together, we obtain:

Theorem: If D is a region of the plane that is simultaneously type I and type II then,

$$\int_C Pdx + Qdy = \iint_D \left(\frac{\partial Q}{\partial x} - \frac{\partial P}{\partial y} \right) dx\,dy.$$

Green' Theorem can easily be extended to any region that can be decomposed into a finite number of regions with are both type I and type II. Such regions we call "nice". Fortunately, most regions are nice. For example, consider the region below.

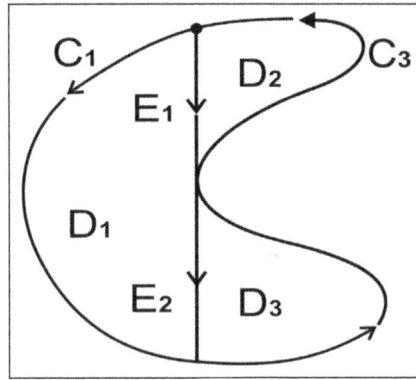

Since D is the union of D_1, D_2 and D_3, we have:

$$\iint_D = \iint_{D_1} + \iint_{D_2} + \iint_{D_3}.$$

Since the regions D_1, D_2, D_3 are all type I and type II and the positively oriented boundaries of D_1, D_2 D_3 are respectively $C_1 - E_2 - E_1$, $E_1 + C_2$, $E_2 + C_2$, we have:

$$\iint_D \left(\frac{\partial D}{\partial x} - \frac{\partial P}{\partial y} \right) dx\,dy = \int_{C_1} P\,dx + Qdy - \int_{E_1} P\,dx + Q\,dy - \int_{E_2} P\,dx + Q\,dy,$$

$$\iint_{D_2} \left(\frac{\partial Q}{\partial x} - \frac{\partial P}{\partial y} \right) dx\,dy = \int_{E_1} P\,dx + Q\,dy + \int_{C_2} P\,dx + Q\,dy,$$

$$\iint_{D_3} \left(\frac{\partial Q}{\partial x} - \frac{\partial P}{\partial y} \right) dx\,dy = \int_{E_2} P\,dx + Q\,dy + \int_{C_2} P\,dx + Q\,dy.$$

Adding these equations, we get,

$$\iint_D \left(\frac{\partial D}{\partial x} - \frac{\partial P}{\partial y} \right) dx\,dy = \int_1 CP\,dx + Qdy + \int_{C_2} P\,dx + Q\,dy + \int_{C_3} P\,dx + Q\,dy = \int_C P\,dx + Q\,dy,$$

where $C = C_1 + C_2 + C_3$ is the positively oriented boundary of D. This yields Green's Theorem for D.

The reader is invited to prove Green's Theorem for the region below using the given decomposition into regions which are type I and type II.

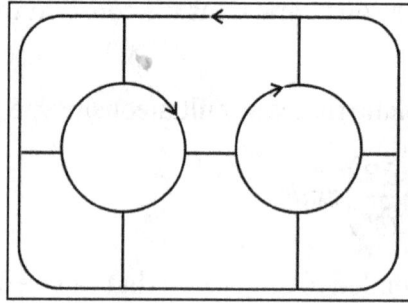

Flux Form of Green's Theorem

Let R be a region for which Green's Theorem holds and let C be the positively oriented boundary of R. For each point P on C let $\vec{T}$ be the unit tangent vector at P and let $\vec{N} = \vec{T} \times \vec{k}$, where $\vec{k}$ is the unit normal to the x, y-plane.

Theorem: If $\vec{F} = P\vec{i} + Q\vec{j}$ is a twice continuously differentiable vector field on R then,

$$\int_C \vec{F} \cdot \vec{N}\, ds = \iint_R \left(\frac{\partial P}{\partial x} + \frac{\partial Q}{\partial y} \right) dx\, dy.$$

Proof: Since $\vec{u} \cdot \left(\vec{v} \times \vec{w} = \left(\vec{w} \times \vec{u} \right) \right) \cdot \vec{v}$, we have,

$$\int_C \vec{F} \cdot \left(\vec{T} \times \vec{k} \right) \cdot ds = \int_C \left(\vec{k} \times \vec{F} \right) \cdot \vec{T}\, ds = \int_C -Q\, dx + P\, dy = \iint_R \left(\frac{\partial P}{\partial x} + \frac{\partial Q}{\partial y} \right) dx\, dy.$$

This theorem is called the flux form of Green's Theorem since,

$$\int_C \vec{F} \cdot \vec{N}\, ds$$

is the flux of $\vec{F}$ across C. The function $\dfrac{\partial P}{\partial x} + \dfrac{\partial Q}{\partial y}$ is called the divergence of the vector field $\vec{F} = P\vec{i} + Q\vec{j}$ and is denoted by div $\left(\vec{F} \right)$. For this reason, Theorem is also called the 2-dimensional Divergence Theorem. Note that, if $\nabla = \dfrac{\partial}{\partial x}\vec{i} + \dfrac{\partial}{\partial y}\vec{j}$, we have:

$$div\left(\vec{F} \right) = \nabla \cdot \vec{F}.$$

The vector field $\nabla \times \vec{F} = \left(\dfrac{\partial Q}{\partial x} - \dfrac{\partial P}{\partial y} \right)\vec{k}$ is the called the curl of the vector field $\vec{F}$ and is also denoted by curl $\left(\vec{F} \right)$. The first form of Green's Theorem can be stated as:

$$\int_C \vec{F} \cdot \vec{T}\, ds = \iint_R curl\left(\vec{F} \right) \cdot \vec{k}\, dx\, dy.$$

These two equivalent forms of Green's Theorem in the plane give rise to two distinct theorems in three dimensions. The usual form of Green's Theorem corresponds to Stokes' Theorem and the flux form of Green's Theorem to Gauss' Theorem, also called the Divergence Theorem.

Green's Theorem can be used to give a physical interpretation of the curl in the case $\vec{F}$ is the velocity field $\vec{v}$ of a flow. If Cr is a circle of radius r with center P, then the average value of the angular velocity $\omega_r = \vec{v} \cdot \vec{T}/r$ on C_r is:

$$\overline{\omega}_r = \frac{1}{2\pi r^2} \int_{C_r} \vec{v} \cdot \vec{T} \, ds.$$

If D_r is the closed disk with boundary C_r, the average value of curl $(\vec{v}) \cdot \vec{k}$ on D_r is:

$$\frac{1}{\pi r^2} \iint_{D_r} curl\left(\vec{v}\right) \cdot \vec{k} \, dx \, dy = \frac{2\overline{v_T}}{r} = 2\overline{\omega}_r.$$

Taking the limit as $r \to 0$, we find that that the angular velocity of the flow around P is,

$$\omega = \lim_{r \to 0} \omega_r = \frac{1}{2} curl\left(\vec{v}\right)(P) \cdot \vec{k}$$

and hence that curl $(\vec{v})(P) = 2\omega \vec{k}$. For this reason, we sometimes denote curl($\vec{v}$) by rot($\vec{v}$). The vector field $\vec{v}$ is said to be irrotational if curl($\vec{v}$) = 0.

Using the flux form of Green's Theorem we can, in the same way, give a physical interpretation of div($\vec{v}$)(P). The flux of $\vec{v}$ across Cr per unit area of D_r is,

$$\frac{1}{\pi r^2} \int_{C_r} \vec{v} \cdot \vec{T} \, ds = \frac{1}{\pi r^2} \iint_{D_r} div\left(\vec{v}\right) dx \, dy = div\left(\vec{v}\right)(Q_r)$$

for some point Q_r in D_r. Taking the limit as r → 0, we find that div($\vec{v}$) (P) measures the rate of change of the quantity of fluid or gas flowing from P per unit area. For this reason, P since called a source if div($\vec{v}$) (P) > 0 and a sink if div($\vec{v}$) (P) < 0. The vector field $\vec{v}$ is said to be incompressible if div ($\vec{v}$) = 0.

Mean Value Theorem

Let f(x) be a continuous function on the interval [a, b] and differentiable on the open interval (a, b). Then there is at least one value c of x in the interval (a, b) such that,

f'(c) = [f(b) - f(a)] /(b - a)

or

f(b) - f(a) = f' (c) (b - a)

In other words, the tangent line to the graph of f at c and the secant through points (a, f(a)) and (b, f(b)) have equal slopes and are therefore parallel.

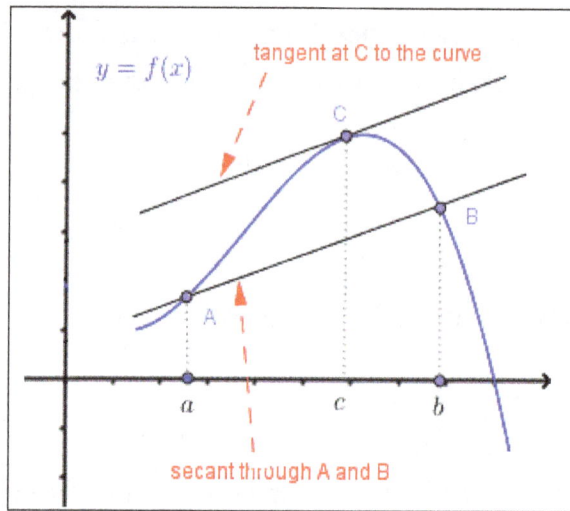

Graphical Meaning of Mean Value Theorem.

Examples on the Applications of the Mean Value Theorem

Example: Use the mean value theorem to find the value c of x in the interval [1 , 5] such that the tangent at the point (c , f(c)) to the of curve $f(x) = -x^2 + 7x - 6$ is parallel to the secant through the points (1 , f(1)) and (5 , f(5)).

Solution to example: The slope of the tangent at point (c , f(c)) is given by, $f'(x)$ where f' is the first derivative.

The slope of the secant through (1 , f(1)) and (5 , f(5)) is given by:

[f(5) - f(1)] /(5 - 1)

For the tangent to be parallel to the secant their slope have to be equal hence:

f' (c) = [f(5) - f(1)] /(5 - 1)

Function f is a polynomial (quadratic) function and is therefore continuous and differentiable of the interval [1, 5] hence the mean value theorem predicts that there is a least one value of x (= c) such that the above equality is true.

The slope of the tangent is given by the value of the first derivative at $x = c$.

The first derivative : f' (x) = - 2 x + 7

slope m_1 of the tangent to the curve at $x = c$ is equal to m_1 = f' (c) = - 2 c + 7.

The slope m_2 of the secant through the points (1, f(1)) and (5, f(5)) is given by:

m_2 = (f(5) - f(1)) / (5 - 1) = (4 - 0) / (4) = 1

m_1 = m_2 gives the equation

$$-2c + 7 = 1$$

$$c = 3$$

Point of tangency at $x = c$ is given by $(3, f(3)) = (3, 6)$:

Equation of tangent:

$$y - 6 = (x - 3)$$

$$y = x + 3$$

In figure are shown the graphs of the given function and the graph of the tangent to the curve of f. The tangent and secant have equal slopes and are therefore parallel.

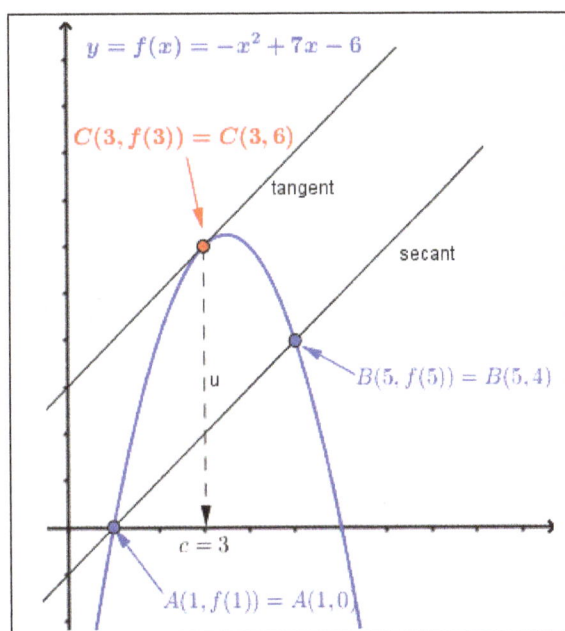

Mean Value Theorem.

There may be more that one value of x ($= c$) that satisfies the mean value theorem below.

Example: Use the mean value theorem to find all values of x in the interval $[0 , 3]$ such that the tangent at the points $(c , f(c))$ to the of curve $f(x) = x^3 - 5x^2 + 7x + 1$ is parallel to the secant through the point $(0, f(0))$ and $(3, f(3))$.

Solution to example: Function f is a polynomial function and is therefore continuous and differentiable of the interval $[1, 3]$ and therefore the mean value theorem predicts that there is at least one value of x ($= c$) such that the tangent to the curve of f at $x = c$ and the secant are parallel and therefore their slopes are equal.

The first derivative : $f'(x) = 3x^2 - 10x + 7$.

The slope m_1 of the tangent at $x = c$ is equal to $m_1 = f'(c) = 3c^2 - 10c + 7$.

The slope m_2 of the secant through the points $(0 , f(0))$ and $(3 , f(3))$.

$$m_2 = (f(3) - f(0)) / (3 - 0) = (4 - 1) / (3 - 0) = 1.$$

For the tangent to the curve at $x = c$ and the secant through $(0, f(0))$ and $(3, f(3))$ to be parallel, their slopes have to be equal.

$$3c^2 - 10c + 7 = 1$$

which may be written as:

$$3c^2 - 10c + 6 = 0$$

Solve using quadratic formulas to obtain two solutions:

$$c_1 = (5 - \sqrt{7})/3 \approx 0.78 \text{ and } c_2 = (5 + \sqrt{7})/3 \approx 2.55$$

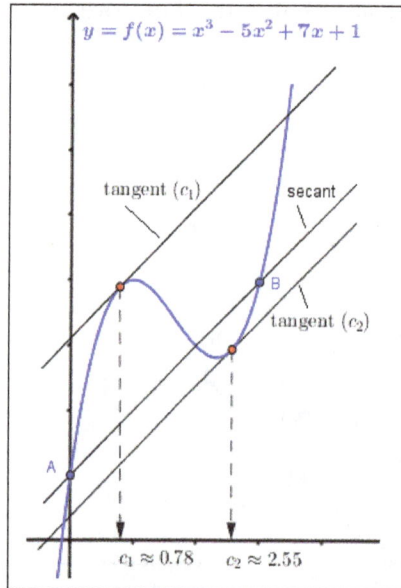

In figure are shown the graphs of the given function and the graph of the two tangents to the curve of f parallel to the secant through the points A(0, f(0)) and B(3, f(3)).

References

- Rolles-theorem: math24.net, Retrieved 20 April, 2019

- Divergence-theorem, triple-integrals-and-surface-integrals, mathematics: mit.edu, Retrieved 26 January, 2019

- Divergence-theorem, Advanced-Calculus: ncku.edu.tw, Retrieved 6 May, 2019

- Gradient-theorem-line-integrals: mathinsight.org, Retrieved 11 April, 2019

- Gradient-theorem-simple-examples: mathinsight.org, Retrieved 4 June, 2019

- Stokes-Theorem: conservapedia.com, Retrieved 10 March, 2019

- Stokes-theorem-examples: mathinsight.org, Retrieved 15 July, 2019

- Stokes-theorem-orientation: mathinsight.orgs, Retrieved 5 August, 2019

- MeanValueTheorem, calculus: analyzemath.com, Retrieved 9 February, 2019

Permissions

All chapters in this book are published with permission under the Creative Commons Attribution Share Alike License or equivalent. Every chapter published in this book has been scrutinized by our experts. Their significance has been extensively debated. The topics covered herein carry significant information for a comprehensive understanding. They may even be implemented as practical applications or may be referred to as a beginning point for further studies.

We would like to thank the editorial team for lending their expertise to make the book truly unique. They have played a crucial role in the development of this book. Without their invaluable contributions this book wouldn't have been possible. They have made vital efforts to compile up to date information on the varied aspects of this subject to make this book a valuable addition to the collection of many professionals and students.

This book was conceptualized with the vision of imparting up-to-date and integrated information in this field. To ensure the same, a matchless editorial board was set up. Every individual on the board went through rigorous rounds of assessment to prove their worth. After which they invested a large part of their time researching and compiling the most relevant data for our readers.

The editorial board has been involved in producing this book since its inception. They have spent rigorous hours researching and exploring the diverse topics which have resulted in the successful publishing of this book. They have passed on their knowledge of decades through this book. To expedite this challenging task, the publisher supported the team at every step. A small team of assistant editors was also appointed to further simplify the editing procedure and attain best results for the readers.

Apart from the editorial board, the designing team has also invested a significant amount of their time in understanding the subject and creating the most relevant covers. They scrutinized every image to scout for the most suitable representation of the subject and create an appropriate cover for the book.

The publishing team has been an ardent support to the editorial, designing and production team. Their endless efforts to recruit the best for this project, has resulted in the accomplishment of this book. They are a veteran in the field of academics and their pool of knowledge is as vast as their experience in printing. Their expertise and guidance has proved useful at every step. Their uncompromising quality standards have made this book an exceptional effort. Their encouragement from time to time has been an inspiration for everyone.

The publisher and the editorial board hope that this book will prove to be a valuable piece of knowledge for students, practitioners and scholars across the globe.

Index

www.ingramcontent.com/pod-product-compliance
Lightning Source LLC
Chambersburg PA
CBHW082014190326
41458CB00010B/3188